The Journal of Modern Craft

Volume 5—Issue 1—March 2012

Editors

Glenn Adamson
Tanya Harrod
Edward S. Cooke, Jr.

Aims and Scope

The Journal of Modern Craft is the first peer-reviewed academic journal to provide an interdisciplinary and international forum in its subject area. It addresses all forms of making that self-consciously set themselves apart from mass production—whether in the making of designed objects, artworks, buildings, or other artefacts.

The journal covers craft in all its historical and contemporary manifestations, from the mid-nineteenth century, when handwork was first consciously framed in opposition to industrialization, through to the present day, when ideas once confined to the "applied arts" have come to seem vital across a huge range of cultural activities. Special emphasis is placed on studio practice, and on the transformations of indigenous forms of craft activity throughout the world. The journal also reviews and analyzes the relevance of craft within new media, folk art, architecture, design, contemporary art and other fields.

The Journal of Modern Craft is the main scholarly voice on the subject of craft, conceived both as an idea and as a field of practice in its own right.

Funded in part by the Windgate Charitable Foundation with support from The Center for Craft, Creativity & Design, whose mission is to advance the understanding of craft by encouraging and supporting research, scholarship and professional development. **www.americanstudiocrafthistory.org.** *The Journal of Modern Craft* gratefully welcomes this and all other suppport that it receives.

Submissions
To submit an article for consideration please contact Glenn Adamson at
g.adamson@vam.ac.uk

Subscription Information
Three issues per volume. One volume per annum.
2012: volume 5

Online
www.bergjournals.com/journalofmoderncraft

By Mail
Berg Publishers
C/o Customer Services
Turpin Distribution
Pegasus Drive
Stratton Business Park
Biggleswade
Bedfordshire SG18 8TQ
UK

By Fax
+44 (0)1767 601640

By Telephone
+44 (0)1767 604951

Subscription Rates
Institutional:
Print and online: 1 year: £184/US$358;
2 year: £295/US$573
(VAT charged if applicable)

Individual:
Print: 1 year: £28/US$54; 2 year: £45/US$87

Full color images available online
Access your electronic subscription through
www.ingentaconnect.com

Reprints for Mailing
Copies of individual articles may be obtained from the publishers at the appropriate fees. For information, write to

Berg Publishers
50 Bedford Square
London WC1B 3DP
UK

Inquiries
Editorial:
Julia Hall, email: julia.hall@bloomsbury.com

Production:
Ken Bruce, email: ken.bruce@bloomsbury.com

Advertising:
Ellie Graves, email: eleanor.graves@bloomsbury.com

Berg Publishers is a member of CrossRef

The Journal of Modern Craft is indexed by Abstracts in Anthropology; ARTbibliographies Modern; Current Contents/Arts and Humanities; Design & Applied Arts Index (DAAI); International Bibliography of Book Reviews of Scholarly Literature in the Humanities and Social Sciences (IBR); International Bibliography of Periodical Literature on the Humanities and Social Sciences (IBZ); ISI Arts and Humanities Citation Index.

ISSN 1749-6772

Typeset by JS Typesetting Ltd, Porthcawl, Mid Glamorgan

The Journal of
Modern Craft

Volume 5—Issue 1—March 2012

Contents

The New Ceramics
A new series from A&C Black

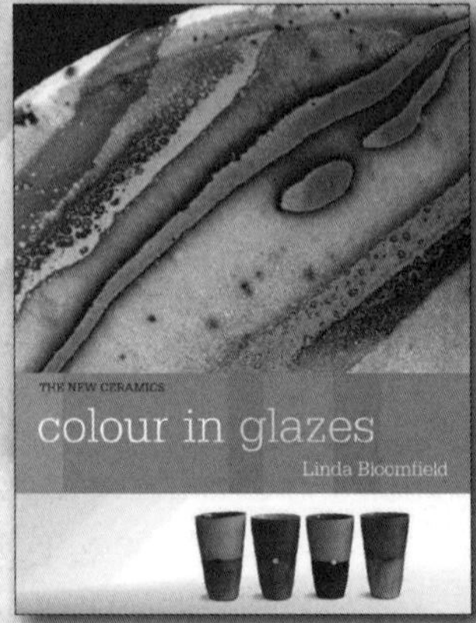

A focused analysis of the potential of ceramic transfer printing as a creative medium.
Jan 2011 / £16.99 pb
9781408113288

A complete guide to achieving a fantastic spectrum of colourful glazes for the studio potter.
Dec 2011 / £16.99 pb
9781408131213

www.acblack.com/visualarts

The Journal of
Modern Craft

Volume 5—Issue 1
March 2012
pp. 5–8

DOI:
10.2752/174967812X13287914145352

Editorial Introduction

The pleasures of craft work are often said to reside in its immediacy: the direct access to materials, the handling of tools, and the sense of accomplishment. Even watching a demonstration in person can be an absorbing experience. Yet texts about craft, including this journal, must necessarily present secondhand the process of making. Language alone simply cannot account for craft's scope of experience. Drawings, paintings, photographs, films, and virtual simulations, all in their own ways, would seem to fill this evident gap, transmitting the reality of skilled work in something closer to its fullness. However, they usually fall short. In the representation of process, such images create a new, different level of material reality, one that needs to be analyzed in its own right.

In this issue we concentrate on the phenomenon of "showing making," a phrase proposed by Dutch scholar Ann-Sophie Lehmann. When welded together, these two verbs suggest the complexity of craft-in-representation, which always involves a dynamic interplay between artisan, artifact, tool, and image. Each of our contributors examines instances of such convergence. Three articles are drawn from a conference held in 2009, which was organized by Lehmann with Nico de Klerk at the Filmmuseum in Amsterdam (EYE), supported by the Meertens Institute (Royal Netherlands Academy of Arts and Sciences). Lehmann's own theoretical overview presents a methodology for the study of "showing making," and then applies it to the example of hand-colored Japanese photographs. She demonstrates the recursive logic of these images, which offer a kinaesthetic pleasure to the viewer while also constructing a self-referential impression of a craft formed at the intersection of traditional painterly skills and new technology.

The papers by Victoria Cain and Irene Steng, also drawn from the 2009 conference, illuminate two other contexts in which photographic images extend and transform the meaning of craft. Cain focuses on an intriguing case study: the preparators who made dioramas at the Natural History Museum in New York City in the earlier 1900s, especially in the 1920s and 1930s. These workers' skills recall those of taxidermists, propmakers, and scientists, but had a specificity and theatricality of their own, which were exploited enthusiastically by the museum in its programmatic and promotional activities. Cain situates staged photographs of the preparators within a broader range of images of craft process circulating in the interwar period. Her argument is that these decades—often thought of as a period fascinated with machines and technology, to the exclusion of handwork—were in fact saturated with such pictures. This widely shared "craftsmanship aesthetic," she writes, offered an ameliorative or reassuring counterpoint to narratives of technological progress that were equally current at the time. Cain's article can be set alongside Ezra Shales's analysis of the Empire State Building (published in our July 2011 issue) as a major contribution to the understanding of modern craft in interwar America, outside the boundaries of the incipient studio movement.

At first, Stengs's article, on the representation of kingship in Thailand, could not seem more different. She shows how the carvers and gilders who make sculptures of the Thai rulers operate in relation to popular photographs. Another example of "showing making" arises in her discussion of live demonstrations that are conducted in markets and temple complexes. This performance of craft takes its place within a diverse image-scape which has as its goal the consolidation of national identity. Perhaps it is only through the unstated relation of these various representational registers that such an impression of unity could be achieved.

Henrietta Lidchi's discussion of Native American jewelers also involves the analysis of a single craft from multiple angles. Historic photographs and live demonstrations again play a role in her account, as do written texts, oral history interviews, and Lidchi's firsthand observations of the Southwestern markets

in which iconic silver and turquoise jewelry is displayed and sold. The article is exemplary in its juxtaposition of past and present, showing how the tools of anthropology can be brought to bear on both history and the present day.

In her manifesto on "showing making," Lehmann alludes to the oft-used phrase "the social life of things," originally formulated by Arjun Appadurai. She insists that this biographical model needs to be extended to include the making of objects, and to this we might add historical precedents—the crafts of the past that make present endeavors possible. A biological or familial metaphor is at the heart of this issue's Statement of Practice by boatmaker Gail McGarva. She has dedicated her life to the replication of open-sea working vessels, vernacular designs carrying strong associations with particular stretches of the British shoreline. McGarva refers to her lovingly made copies as "daughterboats," a way of capturing the generational rhythms of craft succession. Given her interest in such legacies as the taproot of contemporary communities, it is perhaps no surprise that she makes her boats in public and invites others to watch, and even participate in the building process. This is another example of "showing making," this time to the same community that developed and supported the regional product in the first place.

Finally, we include a primary text that is not a description of craft process, but rather a spectator's response. The author is the indomitable Margaret M. Patch, who, despite her relatively advanced years, went on an extraordinary, round-the-world-in-eighty-days-style tour (though it took her a bit longer than that) in the early 1960s.

Her mission was to compile a list of the leading contemporary craft reformers, activists, and developers in advance of the inaugural conference of the World Crafts Council, held in New York City in 1964. As she traveled, Patch paid close attention to cultural differences in practice and attitudes to skill. The previously unpublished text we include here was written early in her journey, and compares the craft cultures of Japan and India. Patch had a high regard for the artisans she found in both places, but was dismayed at the low status of those she encountered in India. This prompted her to reflect on questions of aspiration and recognition that had implications for craft anywhere, including back home in the United States. This is one example of the way that "showing making" can be an invitation to consider one's own act of looking, and hence position in the politics of skill.

The Editors
The Journal of Modern Craft

The Journal of Modern Craft   Volume 5—Issue 1—March 2012, pp. 5–8

**The Journal of
Modern Craft**

Volume 5—Issue 1
March 2012
pp. 9–24

DOI:
10.2752/174967812X13287914145398

Showing Making: On Visual Documentation and Creative Practice

Ann-Sophie Lehmann

Ann-Sophie Lehmann is Associate Professor at the Department for Media and Culture Studies at Utrecht University. Her research develops historical and theoretical perspectives on materials, tools, and practices used in artifact production in old and new media cultures. She is an editorial board member of the *Netherlands Yearbook for History of Art* and co-founder of the Iconology Research Group.

Abstract

This article discusses visual representations of creative practice as source material to study the complex procedures involved in the making of artifacts. After a brief discussion of theoretical approaches to production processes, the genre of *showing making* is introduced as a new line of enquiry. The still and moving images belonging to the genre contain at least four main elements which are extremely useful to the analysis of making. They have an archival function, as they store tacit knowledge about making; an instructional function, in that they enable the acquisition of skills and material knowledge; a participatory function, in the sense that demonstration incites pleasure in the viewer through kinaesthetic identification with the depicted process; and finally a display function, which showcases some but hides other elements of the creative process. These functions can be made instrumental to the study of creative practice, it is finally argued, if the mediated nature of images (which is the result of material processes) is taken into account.

Keywords: creative process, image making, materials, mediation.

Recent theoretical approaches to artifacts in the adjacent fields of anthropology, sociology, archaeology, and material culture studies all agree in one respect: if we want to understand what artifacts mean—in the broadest sense of the word—we have to investigate the complex, dynamic networks in which they are created, used, modified, collected, and destroyed.[1] Artifacts circulating within networks are often described using anthropomorphic metaphors, as in the well-known phrases "the social life of things" (Appadurai 1986), "the everyday life of objects" (Clark 2007), "the material life of things" (Lucchini 2010), and the "life cycle of artifacts" (Bijker 2010; Boradkar 2010). These metaphors have been useful in overcoming subject-object dualisms by emphasizing the idea that things, too, possess agency (e.g. Gell 1998; Latour 2005). Yet this biological analogy, in which things are likened to living organisms, is also misleading because it suggests that things only really exist from the moment that they are "born." And indeed many studies devoted to particular artifacts consider them primarily as finished objects, paying little attention to their becoming or, in keeping with the metaphor, their "prenatal" life.[2] In mammals, the prenatal phase of growth is homeostatic, in principle hidden from view and inaccessible except in particular medical or experimental settings. Therefore, it is usually excluded from the conception of life proper; thus a biography typically starts from birth.

The case is completely different with artifacts. As the very term (with its etymological root of *factum*) denotes, they are not the result of growing but of making, and while growing is a self-regulated, internal process, making is embedded in and accessible to human experience. Of course, everybody knows that things are not alive and it may seem naive to take the metaphor so seriously. Yet the contrast between growing and making makes clear that if the analysis of things ignores processes of production, it fails to acknowledge how the complex interaction between humans, materials, tools, and technologies shapes the possible meanings and usages of the resulting artifact.[3] Understanding making therefore is vital to understanding artifacts.

Making Is Motion

The complexity of making poses a challenge to academic research. Theoretical approaches to what actually happens when a thing is made are only just being developed. A contemporary definition may describe the act of making as a temporary creative unit, fixed in time and place, in which materials, tools, and maker interact. In his book *The Craftsman*, Richard Sennett has described the knowledge resulting from this creative unit as "material consciousness," leaving open whether it resides in the human agent alone or also in the materials involved (2008: 119–46). The archaeologist Lambros Malafouris, in line with Actor-Network Theory, argues that an analysis of making cannot focus on the individual agents involved, but must study what happens *between* makers, materials, and tools. He describes this in-between as material engagement, or the "grey zone where brain, body and culture conflate" (2010: 22), a zone essentially defined by a constant flow of activities, from neurons firing, to hands moving, to materials resisting, and back. Howard Risatti in his *Theory of Craft* gives a similar though more technical definition, when he writes: "Both material and process are essential to craft and must be understood together as the basis of craft technique, a unity of operations centered in functional purpose" (2007: 99). And Glenn Adamson has pinpointed motion as the prime catalyst of this unity, stating that craft really "only exists in motion" (2007). The connection of creative actions through purposeful motion is also paramount in Timothy Ingold's interpretation of making. The anthropologist compares artists and artisans to "itinerant wayfarers" who "make their way through the taskscape as do walkers through the landscape." To follow the paths that they have waded through materials is to understand creativity (2010: 97).

A brief historical perspective on this line of thinking suggests that motion is indeed inherent to theorizing creativity. John Dewey had already stated in *Art as Experience* that every work of art always denotes "a process of doing or making" (1934: 47) and that art theory should therefore expand static and distanced aesthetic contemplation towards including dynamic and performative aspects of artistic production. Dewey envisioned this production as an exchange between two different material spheres. During making, he wrote, "inner material" (everything commonly ascribed to the embodied mind: thought, observation, memories, imagination, emotion) and "outer material" (the physical stuff the artist employs) interact. As "the physical process develops imagination, while imagination is conceived in terms of concrete material" (74–5), this exchange creates a purposeful back and forth movement, which results in an art object.

While the different yet related theoretical approaches briefly assembled here all agree that a theory of making needs to address the unit of agents (makers, materials, tools) and their dynamic interaction, the different

ways of expressing this idea also hint at the greatest obstacle such approaches may meet: how exactly can the interchange between inner and outer materials be grasped? How do we trace the paths that artisans have trodden as they have waded through their "taskscapes"? In other words, how can theoretical descriptions like these, which beautifully capture the complexity of making without necessarily clarifying it, be particularized in detailed studies of creative practice?

How to Study Making

One problem certainly lies in the fact that the complexity of making challenges our most important analytical tool: written language and its essentially linear structure. Richard Sennett has argued that language "is not an adequate 'mirror tool' for the physical movements of the human body," and that only visual experience can take in making fully, and hence "enable our eyes to do the thinking about material things" (2008: 95). Back in the fifteenth century, the authors of instructional manuals warned the reader about the insufficiency of their own medium: in order to learn making one has to see it—ideally by looking over a master's shoulder—and not only read about it (Cennini 1960: 65). For theoretical studies of making, which of course need language, too, the deficiency of textual description is easier to balance when there is a direct access to the act of making and thus a lot of information, as in anthropological fieldwork or other encounters with contemporary process. Past making is exponentially more difficult to study. Sources may be lacking entirely, or else deemed unreliable—such as artists' biographies, which are often unfairly neglected because they are considered merely anecdotal. Manuals and recipe books have only recently been discovered by researchers outside the field of restoration. A close reading of many recipes for the same thing, or a detailed account by a single artisan, may give an impression of the complexity of historic ways of making (see also Lehmann 2008; Smith and Beentjes 2010). Archaeologists have neither direct observation nor textual sources at their disposal and have to extract all information about making from the object itself (see also Bleed 2008). This is not to say that a theoretical analysis of making and its materials is unattainable. It does, however, mean that methods need to be developed, and language carefully employed, so as to match the complexity of making (Elkins 2008; Herzfeld 2007).

Generally speaking, studies of artifact-making can develop along four lines. Two of these are the province of particular academic fields: first, direct observation of process (anthropology, sociology); second, scrutinizing objects for traces of making (archaeology, art history, history of science). In addition, scholars working in any field may analyze textual descriptions of making and—the fourth line of development— engage in reconstruction and re-enactment. These methods have all been explored to a greater or lesser extent by other writers. I would like to add a fifth line of enquiry here, which is to study the visual documentation of making in still or moving images. If direct observation enables "thinking about material things," as Sennett puts it, so too does mediated observation through visual documentation available in drawings, paintings, photographs, films,

animation, computer simulation, etc. Images can capture the complexity and simultaneity of making where words fail to do so. This is well known, of course, and popularizing as well as academic studies of practice are more often than not accompanied by visual material. But here images are typically used as illustrations, as transparent windows that seem to grant direct access to whichever process is discussed, substituting for the direct visual experience. What is new about the approach proposed here is that it aims to turn images into an analytical tool to investigate making by addressing not only what is shown, but how it is shown, how the image acquires the agency to show making and how its own materiality relates to the material process depicted.

Showing Making

The search for visual information about the making of artifacts reveals a distinctive iconography, surprisingly consistent throughout history and across cultures, geographies, and media. Taken together, images of people making something crystallize into an independent genre, the genre of "showing making." People engaged in crafting objects were depicted on Egyptian murals as early as 1400 BC, and Roman and early medieval artists created representations of miniature painters, glassblowers, weavers, and woodworkers. But images of making first became common in the early modern period with the institutionalization of craft within the guild system, which helped to encourage the production of illustrated craft manuals and portraits of artists and artisans. In the eighteenth century, crafting procedures were visually recorded for encyclopedic projects; the nineteenth century saw countless allegorical and topological representations of craftspeople at work, and photography permitted extensive documentation of craft procedures, often their consecutive steps. In the twentieth century, film emerged as an ideal means to record the movements of production, and making became a favorite subject in documentary and educative film as well as television. In the age of new media, the visual representation of practice has multiplied in online tutorials, photo-sharing platforms, and YouTube videos that demonstrate "how to" make every artifact imaginable, from throwing bowls, weaving cloth, or folding origami to making digital artifacts with advanced software (Lehmann 2012).

The persistency and robustness of the genre through the ages—traveling with ease across different visual media and between art and popular culture—suggests that showing making has a basic anthropological function in helping to understand the complex nature of the human capacity to make things. Pictures which show making contain thickly layered information about the social, material, technical, cultural, and aesthetic dimensions of the production of artifacts. While specific groups of images have been studied from some of these perspectives, there has never been an integral approach to the genre as a whole.[4] A critical analysis, with the goal of unpacking the wealth of information these images contain, may be useful to visual studies and art history as it helps to reconcile these fields with the realm of "making and doing," as Dewey called for. The visual analysis of making can also contribute to theories of creative practice as they currently develop in anthropology, sociology, material culture studies, and craft and design studies.

Knowledge, Skill, Pleasure, and Display: The Elements of Showing Making

Showing making has a number of key functions, which are more or less explicitly present in all examples of the genre, and which traverse the media used to depict or record the process at hand. First and foremost, representations of making are visual archives in which information and therefore knowledge *about* making is stored. How is glass shaped and blown (see also O'Connor 2007); how is clay kneaded and thrown on the wheel (see also Lehmann 2009; Malafouris 2010); how is a laptop assembled or a chocolate meringue cake baked? Archiving knowledge can be considered the passive function of the genre.

Second, showing making has an instructional value and can enhance the acquisition of skills for making things, such as in manuals, tutorials, and how-to videos. Even staged representations of making, such as a self-portrait in which a painter shows herself at the easel, or a "making-of" feature about a Hollywood movie, contain elements of instructional value in so far as they are generally truthful to procedures of material production (Lehmann 2006). As conveyor of skill, the image mediates between the domains of implicit and explicit knowledge and can be considered the active function of the genre.

A third purpose of showing making is to evoke pleasure through embodied identification. It is this aspect that makes both the passive, archiving function and the active, instructional function available to the viewer: looking at a film of someone drawing, for instance, generates knowledge about the process we see and it might enhance our own drawing skills. It also creates pleasure because we vicariously experience making the drawing ourselves, activating our capacity for kinaesthetic empathy. In the seventeenth century, René Descartes observed that a blind man extends his sense of touch through his cane, just like our hand seems to feel the tip of the pen on the paper—and not the pen in our fingers—when we write. Descartes concluded that we are able to extend our sense of touch beyond our physical being.[5] But the sense of touch can be extended even further because we can also "feel" the pen in the hand of someone else when we observe writing. Recent neurological research confirms that the physical experience of an action we merely perceive visually is a hard-wired neuronal part of perception. The existence of so-called mirror neurons shows that observing a familiar act triggers the same neurons that would fire if we were to perform the act ourselves. The more experience and knowledge we have of the action we observe, the more neurons fire (Calvo-Merino et al. 2006). This effect can also be observed outside the context of simultaneous observation: when looking at a handwritten letter of the alphabet, our brains replays (as it were) the movement that initially formed it, because our hand and our brain know how the letter was written (Freedberg and Gallese 2007).

This "hand in the brain" not only explains why the observation of the skilled action of others is such an essential element of learning a craft (Marchand 2007). It also elucidates why we enjoy watching artists and artisans even when we do not intend to learn their skills. While it takes talent and years to learn drawing or throwing clay or stone knapping well, we can vicariously share in the immediate experience of skilled actions because we all know a bit about making marks on a surface, shaping soft material, or wielding a tool against a resistant material. Showing making extends this knowledge and allows the onlooker to participate, even if only and literally *second-hand*, in the skilled action of another person.

Finally, showing making can serve to put skillful practice on display in order to claim ownership over a certain technique, to create awe in the viewer, or simply to show off. This is often achieved by showing only certain elements of making. Like watching a magic trick, the viewer witnesses a process but cannot discern how the making actually happens. The display function is paradoxical because representation is employed to mystify rather than to clarify creation.

Showing Making Is Mediated

For the elements outlined here to become effective tools in the study of creative practice, the relation between *making*—the process—and *showing*—the image—needs to be taken into account. This relation is far from a simple one. First of all, drawings, paintings, prints, photographs, films, etc. are mediated objects that answer to specific pictorial conventions. These conventions and styles are shaped by the historical, cultural, and geographical settings in which an image is produced. Furthermore, images of making are only rarely direct transcriptions of process. When capturing these procedures, innumerable decisions are made that depart from strict objectivity, from the technologies used to editorial choices. Images fragment and compress; in showing one thing, they leave out another, and can easily be made to idealize, mystify, or obscure creative processes.[6] Showing making can only be employed for the study of creative practice if the double nature of the image which shows, but also hides, is acknowledged.

This, however, is not achieved by treating images only as representations, as immaterial signs, symbols, or carriers of information in need of interpretation—a dominant tradition in visual and media studies. Such primarily semiotic approaches tend to focus on the surface of images, leaving their material depth out of sight.[7] But images, too, are materially constructed artifacts. Making them involves tools, materials, and practices, which in turn shape style and meaning. Thinking about mediation as a material and not only a representational process introduces another layer to showing making. As well as the material practices represented *in* the image, we must direct our attention to the materiality *of* the image. It follows that in order to understand making through visual representation, processes of artifact making have to be studied in conjunction with processes of image making.

Coloring Lanterns, Umbrellas, and Photographs

I would like to demonstrate the approach outlined here by examining some photographs by the Japanese photographer Enami Nobukuni, also known as T. Enami (1859–1929). Enami had a large studio in Yokohama and produced a vast number of albumen photographs, lantern slides, and stereo views during the Meiji and Taisho periods.[8] Among the most prominent genres in early Japanese photography, next to landscape scenes and portraits, are what David Odo has called "occupational scenes"

(2008). These show typical "Japanese" views like geisha drinking tea or Sumo wrestlers, but also include people at work, often engaged in crafting everyday objects. These images were attractive to tourists as well as to Japanese collectors (Odo 2008: 10–12). Enami's oeuvre includes a large number of such images.[9] Within this group a certain subset can be distinguished, showing both male and female artisans engaged in coloring common traditional artifacts such as screens, prints, fans, umbrellas, and lanterns. The photograph of the lantern painters illustrated here (Figure 1) was taken around the early 1890s. Of the four elements of showing making, it addresses the archival function of documentation as well as the display function most clearly. The frontal composition and the arrangement of the props point to a studio photograph. The fact that this is a staged image is confirmed by another photograph showing an umbrella painter, who is sitting under the same line of hanging lanterns and taking refreshments from the same teapot. In fact, he appears to be the very same

Fig 1 T. Enami, *The Lantern Painters*, hand-tinted lantern slide, *c.*1892–5. Reproduced with kind permission from Rob Oechsle, t-enami.org.

Fig 2 T. Enami, *The Umbrella Painter*, hand-tinted lantern slide, *c.*1892–5. Reproduced with kind permission from Rob Oechsle, t-enami.org.

man (Figure 2). A later Enami photograph, showing two boys painting lanterns, gives a more immediate impression of the process (Figure 3). The deep perspective, the wooden floor, lanterns, bamboo and brushes piled in the background, the paint bucket, and the fact that the boys have dirty hands and have spread a newspaper on the ground to protect the paint from spilling, suggest that

Fig 3 T. Enami, *Boys Painting Lanterns*, hand-tinted lantern slide, *c.*1905–20. Reproduced with kind permission from Rob Oechsle, t-enami.org.

this might be a view of a real workshop. Not yet instructional, the picture nevertheless conveys more accurate information about the process and the skills involved (the second function of showing making), and also appeals to the "hand in the brain," which can relate to the large brush and imagine the movements that brought about the colorful patterns on the lanterns. In fact, in Japanese calligraphy, the process is as much valued as the result, and the ability to recreate brush movements from written signs and decorative patterns in the mind was probably high, adding a further kinaesthetic layer to the image (Nakamura 2007). The immediacy is also due to technical developments in photography that enabled a higher sensitivity to light and shorter shutter times, allowing the camera to move into new spaces.[10]

The most fascinating kinship between the processes on display and the material process of image making itself, however, lies in the coloring. Most of Enami's photographs were hand-colored and famous for being so. Techniques of hand-coloring, though invented in Europe, were brought to perfection in Japan, where traditional skills were applied to these new pictorial objects (Henisch and Henisch 1996; Lacoste 2010). In one of the first large photographic projects to issue a carefully staged image of Japan to the Western world, *Captain Frank Brinkley's Japan: Described and Illustrated by the Japanese* (published between 1897 and 1898; see also Hockley 2011), the ten folio-sized volumes of essays were accompanied by 260 hand-tinted albumen photographs. In Japan, 350 colorists worked on the project, including Enami's studio (Oechsle 2007–11). A black and white photograph shows employees in the workshop in Yokohama engaged in this work, an image that is certainly staged (given the careful display of different formats and phases of the coloring process), yet probably directed at insiders to the trade (Figure 4).[11] But scenes of coloring photographs also became a subject within collectable albums, extending the iconography of decorating lanterns, umbrellas, fans, or screens. These photographs were colored themselves, depicting the process they had undergone, and they also emphasized the pleasure of looking at these new and vividly colored artifacts, as is clear in an Enami photograph that shows boys engaged in coloring and looking at pictures (Figure 5).[12] Looking at photographs became an occupational motif

in its own right, showing, for instance, geisha contemplating pictures, either directly or through a stereoscope. In these scenes, the coloring process is not represented but is cleverly emphasized by coloring everything in the image but the depicted photographs.[13]

All of the pictures briefly described here reflect on the production and materiality of both the emerging medium of photography and the familiar medium of painting, as well as the new liaison they entered. The ensuing multilayered relation between showing and making not only reveals the appreciation of traditional crafts alongside a related, new craft technique. It also shows how coloring played an important role in linking

Fig 4 Unknown, *Hand-tinting of Photographs in the Studio of T. Enami*, half-tone image published in England in 1904, reproduced from albumen print, c.1895–7. Reproduced with kind permission from Rob Oechsle, t-enami.org.

Fig 5 After T. Enami, *Japanese Boys Coloring Pictures and Using Stereoscope*, lithograph from a photograph by T. Enami, 1906–10. Reproduced with kind permission from Rob Oechsle, t-enami.org.

photography to familiar expressions of visual culture and helped to appropriate the new medium as Japanese. Finally, it stresses the cultural value attached to the art of coloring in general: an art that adorns manufactured objects, bringing images made with a mechanical device to life with a manual technique.

Conclusion

The display of hand-coloring photographs *in* hand-colored photographs allows for a particularly distinct demonstration of the relation between the material practices represented in the image and the materiality of the image. This material bond, however, is also relevant in less obviously related media, such as video and clay, for instance.

Extremely corporeal craft techniques like throwing and kneading clay can only be captured fully in moving images. In the many YouTube video tutorials that Simon Leach, grandson of Bernard Leach, has embedded on his website (www.simonleachpottery. com), the dexterity of handling clay stands in stark contrast to the amateur use of the recording medium. In a video that demonstrates the Japanese *kikuneri* wedging technique, Leach walks up to the running camera and adjusts it to the table he is kneading on (Lehmann 2009). The resulting close-up cuts off his head and shows only his hands and arms engaged with the clay while he is talking outside the frame (Figures 6a and 6b). Yet it is exactly this casual mode of recording, done by Leach himself, that helps the viewer identify with the material engagement documented here. By creating an almost too close visual contact with Leach's body and the clay, the video

(a)

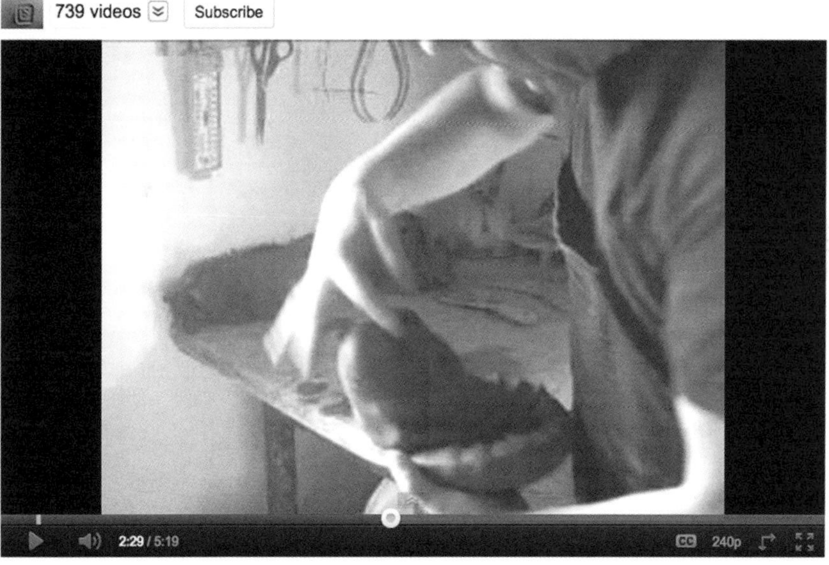

(b)

Fig 6 (a) and (b) Stills from Simon Leach, *Tips on Kneading Clay*, YouTube Video, uploaded June 20, 2007 (http://www.youtube.com/watch?v=xbo1DEdngwY). Reproduced with kind permission from Simon Leach.

images—a bit shaky and not quite in focus—create a distinct impression of the force and smoothness of the kneading motion. As a consequence, the "hand in the brain" is triggered and the demonstration of Leach's wedging becomes very effective.[14]

Like the hand-colored Japanese photographs, the video images of kneading clay can better be understood if their own materiality is considered alongside the material processes they show. Such a combined study of the imaging of craft and the crafting of images opens up new perspectives for media and craft studies—not only because it can tell us about the cultural, social, and material conditions of making, but because it reminds us that images are an integral part of making in their own right.

Notes

1 For an overview of recent theoretical concepts in these and related fields, see Boradkar 2010: 17–43; Knappett and Malafouris 2010.

2 In Science and Technology Studies and Design Studies there is a strong focus on the political, social, and economic aspects involved in the production of things in industrial settings, yet little attention is paid to the actual procedures of making themselves (Boradkar 2010, esp. 75–101).

3 Igor Kopytoff's article "The Cultural Biography of Things" paved the way for a biographical approach to objects. It mentions the initial phase of making as relevant but does not include it in the analysis (Kopytoff 1986).

4 The majority of studies is devoted to pictures of the painter at work; an overview is given in Kleinert 2006. The artist's documentary also has a number of studies devoted to it; see Hayward 1998 and Walker 1993. There are a couple of articles on "making-of" features in film (Hediger 2005; Hight 2005) and on instructional pictures (Gombrich 1999; Lopes 2004).

5 See the Sixth Discourse of René Descartes, *Diotrique* (1637). For a modern interpretation of the phenomenon using the example of chalk as an extension of our hand, see Ihde 1986.

6 This mechanism was the subject of the follow-up conference to "Showing Making": "Hiding Making—Showing Creation. Strategies in Artistic Practice from the 19th to the 21st Centuries," Teylers Museum, Haarlem/Rijksakademie Amsterdam, January 7–8, 2011.

7 For an extended critique of the dominance of immateriality in visual culture, see Ann-Sophie Lehmann, "Das Medium als Mediator. Eine Materialtheorie für Öl(bilder)," *Zeitschrift für Ästhetik und Allgemeine Kunstwissenschaft*, 57 (2012).

8 The oeuvre of Enami has only recently been fully rediscovered, documented, and made available by Rob Oechsle on http://www.t-enami.org/. Under the pseudonym Okinawa Soba, Oechsle maintains a photostream on Flickr with high-quality digitized photographs by T. Enami (http://www.flickr.com/photos/24443965@N08/). See also Bennett 2006.

9 Oechsle has assembled a set of occupational images by Enami on Flickr: "At Work in Old Japan" (http://www. flickr.com/photos/24443965@N08/ sets/72157605714378115/with/2456262266/).

10 Before, craft scenes would also be staged outside, where improvised outdoor studios would allow for enough light and the display of messier crafts. These photographs are called *shajo*, referring to improvised outdoor theater stages. See Clark Worswick, *Japan: Photographs, 1854–1905* (New York: Random, 1973).

11 On photographic studio practice in Japan, see Fraser 2009.

12 For a view of women coloring photographs see the Metadate Database of Japanese Old Photographs in Bakumatsu-Meiji Period, Nagasaki University Library (http://hikoma.lb.nagasaki-u. ac.jp/en/target.php?id=2366).

13 "Geisha Looking at Photographs," assembled by Rob Oechsle, alias Okinawa Soba, on Flickr (http://www.flickr.com/photos/24443965@N08/sets/72157604139392795/with/2339168371/).

14 Vivian Sobchak and Laura Marks describe the phenomenon of bodily identification with moving images as "haptic visuality"; see Vivian Sobchack, *Carnal Thoughts: Embodiment and Moving Image Culture* (Berkeley, CA: University of California Press, 2004), Laura Marks, "Haptic Visuality: Touching with the Eyes," *Frameworks*, 02 (2004): 79–82.

References

Adamson, Glenn. 2007. *Thinking through Craft.* Oxford: Berg.

Appadurai, Arjun. 1986. *The Social Life of Things. Commodities in Cultural Perspective.* Cambridge: Cambridge University Press.

Bennett, Terry. 2006. *Photography in Japan 1853–1912.* Tokyo: Tuttle Publishing.

Bijker, Wiebe E. 2010. "How Is Technology Made?—That Is the Question!" *Cambridge Journal of Economics*, 34: 63–76.

Bleed, Peter. 2008. "Skill Matters." *Journal of Archaeological Method and Theory*, 15: 154–66.

Boradkar, Prasad. 2010. *Designing Things: A Critical Introduction to the Culture of Objects.* Oxford: Berg.

Calvo-Merino, Beatriz *et al.* 2006. "Seeing or Doing? Influence of Visual and Motor Familiarity in Action Observation." *Current Biology*, 16: 1905–10.

Cennini, Cennino d'Andrea. 1960. *The Craftsman's Handbook [Il libro dell'arte].* Translated and edited by Daniel V. Thompson, Jr. New York: Dover.

Clark, Laurie Beth. 2007. Art Project, "The Everyday Life of Things." Art Department, University of Wisconsin-Madison (http://www.everydaylifeofobjects.net).

Dewey, John. 1934. *Art as Experience.* New York: Capricorn Books.

Elkins, James. 2008. "On Some Limits of Materiality in Art History." *31: Das Magazin des Instituts für Theorie*, 12: 25–30. Special Issue, *Taktilität: Sinneserfahrung als Grenzerfahrung*, edited by Stefan Neuner and Julia Gelshorn.

Fraser, Karin. 2009. "Studio Practices in Early Japanese Photography: The Tomishige Archive." *History of Photography*, 33(2): 132–45.

Freedberg, David, and Gallese, Vittorio. 2007. "Motion, Emotion and Empathy in Esthetic Experience." *Trends in Cognitive Sciences*, 11(5): 197–203.

Gell, Alfred. 1998. *Art and Agency: An Anthropological Theory.* Oxford: Clarendon Press.

Gombrich, Ernst H. 1999. "Pictorial Instructions." In Ernst Gombrich, *The Uses of Images: Studies in the Social Function of Art and Visual Communication*, pp. 226–39. London: Phaidon.

Harris, Mark (ed.). 2007. *Ways of Knowing: New Approaches in the Anthropology of Experience and Learning.* Oxford: Berghahn.

Hayward, Philip (ed.). 1998. *Picture This: Media Representations of Visual Art and Artists.* London: John Libbey.

Hediger, Vinzenz. 2005. "Spaß an harter Arbeit. Der Making-of-Film." In Vinzenz Hediger and Patrick Vonderau (eds), *Demnächst in Ihrem Kino. Grundlagen der Filmwerbung und Filmvermarktung*, pp. 332–41. Marburg: Schüren.

Henisch, H.K., and Henisch, B. 1996. *The Painted Photograph 1839–1914: Origins, Techniques, Aspirations.* Philadelphia: The Pennsylvania State University Press.

Herzfeld, Michael. 2007. "Deskilling, 'Dumbing Down,' and the Auditing of Knowledge in the Practical Mastery of Artisans and Academics: An Ethnographer's Response to a Global Problem." In Mark Harris (ed.), *Ways of Knowing: New Approaches in the Anthropology of Experience and Learning*, pp. 91–111. Oxford: Berghahn.

Hight, Craig. 2005. "Making-of Documentaries on DVD: The Lord of the Rings Trilogy and Special Editions." *The Velvet Light Trap*, 56(3): 5–17.

Hockley, Allen. 2011. Globetrotters' Japan: People. "Foreigners on the Tourist Circuit in Meiji Japan." MIT Vizualizing Cultures Project (http://ocw.mit.edu/ans7870/21f/21f.027/gt_japan_people/ga1_essay01.html).

Ihde, Don. 1986. *Experimental Phenomenology: An Introduction.* New York: New York University Press.

Ingold, Tim. 2010. "The Textility of Making." *Cambridge Journal of Economics*, 34: 91–102.

Kleinert, Katja. 2006. *Atelierdarstellungen in der niederländischen Genremalerei des 17. Jahrhunderts: realistisches Abbild oder glaubwürdiger Schein?* Petersberg: Michael Imhof.

Knappett, Carl, and Lambros Malafouris. 2010. "Material and Nonhuman Agency: An Introduction." In Carl Knappett and Lambros Malafouris (eds), *Material Agency: Towards a Non-anthropocentric Approach*, pp. ix–xix. New York: Springer.

Kopytoff, Igor. 1986. "The Cultural Biography of Things: Commoditization as Process." In Arjun Appadurai (ed.), *The Social Life of Things: Commodities in Cultural Perspective*, pp. 64–91. Cambridge: Cambridge University Press.

Lacoste, Anne. 2010. *Felice Beato: A Photographer on the Eastern Road.* Los Angeles, CA: Getty Publications.

Latour, Bruno. 2005. *Reassembling the Social: An Introduction to Actor-Network-Theory.* Oxford: Oxford University Press.

Lehmann, Ann-Sophie. 2006. "Wat de hand weet. De kunstenaar aan het werk." In Mariette Haveman *et al.* (eds), *Ateliergeheimen*, pp. 120–41. Amsterdam, Zutphen: Kunst en Schrijven.

Lehmann, Ann-Sophie. 2008. "Fleshing Out the Body: The 'Colours of the Naked' in Dutch Art Theory and Workshop Practice 1400–1600." In Ann-Sophie Lehmann and Herman Roodenburg (eds), *Body and Embodiment in Netherlandish Art* (Nederlands Kunsthistorisch Jaarboek/Netherlands Yearbook for the History of Art, 58), pp. 108–31. Zwolle: Waanders.

Lehmann, Ann-Sophie. 2009. "Wedging, Throwing, Dipping and Dragging—How Motions, Tools and

Materials Make Art." In Barbara Baert and Trees de Mits (eds), *Folded Stones*, pp. 41–60. Leuven: Acco.

Lehmann, Ann-Sophie. 2012. "Bügeln, Töpfern, Zeichnen, oder: How To YouTube." *Hamburger Hefte zur Medienkultur*, 12: 121–34.

Lopes, Dominic McIver. 2004. "Directive Pictures." *The Journal of Aesthetics and Art Criticism*, 62(2): 189–96.

Lucchini, Francesco. 2010. "The Material Life of Things Project." Courtauld Institute, London (http://www.courtauld.ac.uk/researchforum/projects/materiallifeofthings.shtml).

Malafouris, Lambros. 2010. "At the Potter's Wheel." In Carl Knappett and Lambros Malafouris (eds), *Material Agency: Towards a Non-anthropocentric Approach*, pp. 19–38. New York: Springer.

Marchand, Trevor. 2007. "Crafting Knowledge: The Role of 'Parsing and Production' in the Communication of Skill-based Knowledge among Masons." In Mark Harris (ed.), *Ways of Knowing: New Approaches in the Anthropology of Experience and Learning*, pp. 181–202. Oxford: Berghahn.

Nakamura, Fuyubi. 2007. "Creating or Performing Words? Observations on Contemporary Japanese Calligraphy." In Elisabeth Hallam and Tim Ingold (eds), *Creativity and Cultural Improvisation*, pp. 79–98. Oxford: Berg.

O'Connor, Erin. 2007. "Embodied Knowledge in Glassblowing: the Experience of Meaning and the Struggle Towards Proficiency." *Sociological Review*, 55: 126–41.

Odo, David. 2008. "Unknown Japan: Reconsidering 19th-century Photographs." *Rijksmuseum Studies in Photography*, vol. 4. Amsterdam: Rijksmuseum.

Oechsle, Rob. 2007–2011. Online Repository T-enami.org.

Risatti, Howard. 2007. *A Theory of Craft. Function and Aesthetic Expression.* Chapel Hill, NC: The University of North Carolina Press.

Sennett, Richard. 2008. *The Craftsman.* London: Allen Lane.

Smith, Pamela and Beentjes, Tonny. 2010. "Nature and Art, Making and Knowing: Reconstructing Sixteenth-century Life Casting Techniques." *Renaissance Quarterly*, 63: 128–79.

Walker, John A. 1993. *Art and Artists on the Screen.* Manchester, New York: Manchester University Press.

The Journal of
Modern Craft

Volume 5—Issue 1
March 2012
pp. 25–50

DOI:
10.2752/174967812X13287914145433

The Craftsmanship Aesthetic: Showing Making at the American Museum of Natural History, 1910–45

Victoria E.M. Cain

Victoria Cain is an Assistant Professor/Faculty Fellow of
Museum Studies at New York University, where she studies
the history of visual culture, science and technology, and
museums. She has been awarded fellowships from the
American Academy of Arts and Sciences, the Spencer
Foundation, and the Mellon Foundation, and has published
a number of book chapters and articles, most recently in
the *Journal of Visual Culture* and *Science in Context*. She is the
co-author, with Karen Rader, of a forthcoming book on the
history of display in twentieth-century museums of science
and natural history, and she is currently working on a history
of pictures and technology in American education.

Abstract

The rise of large corporations and unskilled labor in
the early twentieth century conspired to separate many
Americans from knowledge of and pleasure in their
work, and the bewildering specialization of production
methods and professional knowledge only intensified this
sense of alienation. This article argues that a new popular
aesthetic, which celebrated skilled craftsmen laboring
within the exotic provinces of specialized production,
emerged as a result. This aesthetic of craftsmanship was
simultaneously diverting and reassuring, providing viewers
with the visual entertainment associated with tourism
and consumption while connecting them to craftsmen
absorbed in meaningful work. This essay explains how and
why this "craftsmanship aesthetic" flourished, using the
example of photographs of exhibit making at the American

Museum of Natural History between the 1910s and the 1940s. It examines the part this aesthetic played in the museum's public relations campaigns and institutional politics, and explores the ways that the museum's image makers used it to improve their standing within and beyond the museum's halls.

Keywords: Craftsmanship, taxidermy, museum, industrialization, photography, dioramas.

In 1926, a temporary exhibition at the Newark Museum on the production of celluloid, a cheap and easily molded plastic used as a substitute for ivory, attracted attention from unexpected quarters. At lunch hour, groups of workers trundled into the exhibition, "staying only a short time, but poring over the tables and cases with great interest and having much to talk about," reported the *New York Times*. After inquiring, museum staff discovered their visitors were employees in a nearby celluloid factory. According to the *Times*, it was the first time the workers had seen and understood the creation of celluloid "as a whole; they had known only that part of its creation which they handled."[1] The staff members of the Newark Museum were delighted, for, as historian Nicolas Maffei has pointed out, this was precisely what they had hoped would happen. Throughout the decade, in an effort to allow workers to take pride in their role in the production of goods, the museum hosted frequent and successful exhibits featuring the city's manufacturing and craft work, keeping its doors open after the factory whistles blew at night in order to allow the intended

audience to learn how various objects were made, and, often, to watch skilled makers demonstrating their craft.[2]

The Newark factory workers' estrangement from the process of making goods and the museum's efforts to ameliorate this problem embodied, in a single small room, the era's sprawling preoccupation with modern Americans' changing relationship to the production of goods.[3] The ways in which assembly lines, the massive scale of industrial production, and increasing specialization of tasks affected labor aroused both fascination and fear among Americans in the interwar period.[4] Alienation from the production of goods was further intensified by the era's burgeoning consumer culture.[5] Retailers, advertisers, designers, and other engineers of consumer desire took great care to keep processes of production out of sight. They masked the realities of making goods behind a carefully constructed scrim of fantasy, and often attempted to eliminate consumers' thoughts of production entirely.[6]

As science, technology, and psychology transformed early twentieth-century systems of manufacturing, the production of goods became tremendously difficult to visualize or even comprehend, radically changing Americans' relationship to making. In this disorienting age, even commonplace objects—a celluloid collar, for instance—could seem marvelous, for the origins of objects were increasingly obscure, and it was increasingly difficult to draw a straight line between producer and object. Americans were further dissociated from the production of goods as a result of the rise of a powerful complex of institutions and images designed to promote mass consumption. Retailers and advertisers did their best to make the new glut of goods seem visually enticing, obfuscating or distorting histories of production as they did so. Even as they fantasized about newly necessary comforts, Americans began to view these objects of desire with skepticism. As art historian Michael Leja writes, "to function successfully, even to survive" in the early twentieth-century United States, "every participant in the new mass culture, every beneficiary of modern science and technology … had to process visual experience with some measure of suspicion, caution, and guile."[7] Americans became ever more skeptical of the surface appearance of material objects, and ever more fascinated by the sight of their production.[8]

This article argues that these cultural and economic conditions resulted in what I call the "craftsmanship aesthetic." In this context, the term "craftsmanship" refers to skills, abilities, or techniques acquired and burnished over time, while the "craftsmanship aesthetic" refers to the interest early twentieth-century Americans took in contemplating the accomplished making of things, by artisans, artists, or other skilled workers.[9] The aesthetic celebrated the satisfactions of labor, placing human presence squarely at the center of its representation. It exalted the relationship between worker, skill, and tool. The sight of craftspeople gradually, lovingly shaping material into useful or impressive objects resonated with a public struggling with the new dynamics of scientifically managed corporations. It also afforded backstage pleasure, for such representations allowed Americans to glimpse highly specialized sites and processes of production that

were usually inaccessible. Like so many popular aesthetics, the vogue for images of craftsmanship both reassured and stimulated: it simultaneously connected viewers to artisans who were calmly, deeply absorbed in meaningful work, while providing them with visual diversions that could take the form of humor, spectacle, or marvel. Thanks to mass commercial culture, the aesthetic thrived between 1910 and 1945, for its appeal depended on viewers' increasing distance from production as well as an overwhelming desire to look backstage, at what lay beyond the false fronts of shops, shows, and other sites of consumption. And it proved a draw at fairs, department stores, and museums, as well as in the popular press.

Though the craftsmanship aesthetic surfaced in many forms, from print advertising in popular magazines to film shorts to World's Fair displays, the extensive collection of photographs of exhibit making at the American Museum of Natural History (AMNH) provides an excellent starting point from which to sketch out the broad outlines of this phenomenon. The production of natural history museum displays in the early twentieth century fascinated contemporary visitors, and museum administrators capitalized upon this interest. As a result, between the 1910s and the 1940s, the AMNH's photographers produced hundreds of images of expert craftsmen creating these displays. These were circulated among museum visitors and members, as well as in the popular press. The museum's carefully preserved archive of these photographs allows us to trace, through a coherent body of images, the gradual emergence of the craftsmanship aesthetic, as well as some of its permutations and functions, amidst the visual

The Journal of Modern Craft   Volume 5—Issue 1—March 2012, pp. 25–50

culture of production in a single institution during the first half of the twentieth century.

The Emergence of the Craftsmanship Aesthetic

The pleasure that early twentieth-century Americans took in contemplating the skillful making of goods had deep and tangled roots. One of its best-articulated antecedents was a mid-nineteenth-century phenomenon that cultural historian Neil Harris has described as the "operational aesthetic." Americans in the 1840s and 1850s adored examining complicated technologies and well-crafted humbugs, for they "equated beauty with information and technique."[10] The proliferation of awe-inspiring technology and the broader cultural anxieties about visual deception in these decades only furthered the national fascination with how things were made.[11] Amateur expertise and auto-didacticism still prevailed in the United States, and the operational aesthetic of this era frequently assumed the viewer could replicate, contribute to, or, at the very least, comprehend the process under consideration. In novels and newspaper articles, lyceum lectures and personal conversations, antebellum Americans explicated all manner of manufacturing and technology. They parsed everything from the making of olive oil soap to the functioning of the era's great sailing ships to the construction of P.T. Barnum's infamous Feejee Mermaid, reveling in lengthy discussions of the many divergences between appearance and construction or function.

Americans' delight in inspecting artful deceptions or complex technologies persisted, but by the 1910s and 1920s, rather than imagining themselves as potential participants in such productions, Americans often found themselves in the more limited role of consumers. They could still observe, discuss, and enjoy the sophisticated creations they saw, but no longer assumed they could reproduce the phenomena themselves, or even the goods they regularly used or desired.[12] What's more, the increasingly specialized nature of production in these decades and the explosion of scientific knowledge and expertise meant that lay audiences could not always understand the processes by which objects were made.[13] Harris notes that, in the middle of the nineteenth century, avidly democratic Americans had believed that diligent observation and common sense were enough to grasp most aspects of technology and natural science, assuming that "any problem could be expressed clearly, concisely, and comprehensibly enough for the ordinary man to resolve it"; by 1900, however, Henry Adams argued that an average American "could no longer understand the problem," much less solve it.[14]

Knowledge of skilled production also eroded in more intimate settings. In their study of the changing culture of Muncie, Indiana, sociologists Robert and Helen Lynd noted that, between 1890 and the 1920s, leisure had become "more a thing of passive knowledge than of creative employment."[15] Rather than singing or making art, citizens "cease[d] to call forth so much active, constructive ingenuity," and contented themselves with listening to or looking at accomplished renditions of what they could make themselves.[16] Less exalted forms of making—cooking, sewing, and other kinds of domestic handicraft—underwent a similar shift, as Muncie residents increasingly purchased simple, prepared foods, ready-made dresses, and mass-produced goods. Most acknowledged the resulting loss of traditional skills in the face of modern pressures and pleasures. "The modern housewife has lost the art of cooking," as one butcher observed. "Folks today want to eat in a hurry and get out in the car." [17]

Partly as a result of these shifts, a new, though related, aesthetic emerged alongside Americans' ongoing pleasure in understanding and discussing the production of goods. Americans increasingly demonstrated interest in contemplating the makers of goods, not just the making of them. Skilled creators became objects of fascination, and images and descriptions of craftsmen began to appear among depictions of ingenious processes and sublime technology. Like the operational aesthetic of the mid-nineteenth century, what I call "the craftsmanship aesthetic" provided glimpses of the production of the tidal wave of goods that flooded the early twentieth century. Yet at its heart lay the vision of comfortingly competent master craftsmen, not the colder diversions of clever design.

Portraits of accomplished craftsmen at work had long been staples of the nation's iconography, and such images proliferated in the early twentieth century. A visual formula characterized these works: the craftsman or artisan was depicted hard at work or displaying the product of his labor. Rolled-up sleeves, a work apron, and sure hands indicated his artisanal status. [18] Tools of the trade—"bellows-blowers, hammers, and all the et-ceteras of the shop," as blacksmith Pat Lyon, one of the first and most famous subjects of such portraits, described them—rounded out these pictures of honest labor

and moral righteousness (Figure 1). [19] Such images generally presented the artisan working alone, or, occasionally, in quiet concert with a colleague or an apprentice.

Such portraits took on new resonance because of Americans' changed relationship to making. [20] The deskilling of labor in many industries conspired to separate many Americans from knowledge of and pleasure in their work, especially those involved in the manufacture of mass-produced goods on a factory floor. The bewildering specialization of production methods, the proliferation of professional knowledge, and the separation of sites of production from the places of consumption only intensified this sense of

Fig 1 John Neagle, *Pat Lyon at the Forge*, 1826–7. Photograph © 2012, Museum of Fine Arts, Boston.

alienation from the production of objects, for it became difficult for Americans to envision how and by whom goods were made. Though they might not have recognized the names of John Cotton Dana, Ralph Adams Cram, Oscar Lowell Triggs, or even John Ruskin and William Morris, many Americans agreed with these reformers' warnings that the deskilling of workers and the fragmentation of production and consumption were dangerous trends, which would gradually undermine individual morality, cultural advancement, and political democracy.

Exaltations of labor coursed through interwar visual culture in the United States, and the craftsmanship aesthetic was certainly part of this current, but it distinguished itself by featuring skilled artisans in the process of creating, often situating them in the midst of otherwise inaccessible sites and processes of production. The craftsmanship aesthetic most frequently featured two different categories of labor. First, the aesthetic often

glorified people involved in traditional forms of craft endangered by modernity: Americans engaged in the kind of labor that mass production was nudging aside, as well as "primitives": indigenous people involved in native crafts (Figure 2). Yet the aesthetic was not inherently elegiac. It also encompassed accomplished creators of more modern goods toiling amidst the equally exotic settings of industrial, technological, or highly specialized production, a juxtaposition that served to reassure consumers that traditional values of labor remained the same, even in a transformed workplace (Figure 3).

The craftsmanship aesthetic was remarkably malleable, corresponding to a host of social theories popular in the first four decades of the twentieth century. It mirrored strains of anti-modernism,

Fig 3 This photograph, titled "Workmanship," was taken by Westinghouse Electric in 1937. The original caption, written by the Science Service news syndicate, read: "A skilled workman hand-finishing the collector rings of a special 150,000 ampere unipolar generator, built in Westinghouse Electric shops." Science Service Historical Images Collection, Courtesy of Westinghouse. Photograph © 1998–2010 Smithsonian Institution.

Fig 2 Edward S. Curtis, "Untitled." Nampeyo decorating pottery, c. 1900. Library of Congress, Prints and Photographs Division, LC-USZ62-48396.

allowing Americans to meditate on a time when labor practices had been simpler, and the transformation of material more comprehensible.[21] But the aesthetic was not always a visual jeremiad. It could also fold itself into a Spencerian vision of progress or feed into colonialist satisfaction. Visitors to World's Fairs and western reservations, for instance, amused themselves by watching skilled "primitives" engaged in such mysterious crafts as moccasin beading, blanket making, or cedar-bark weaving.[22] Exhibitions like these aimed to entertain while reminding Americans just how far contemporary industrial methods had outpaced the laborious handwork of less civilized cultures. Progressive reformers also made use of this aesthetic, for such representations of the close connection between head and hand, a union increasingly jeopardized in the twentieth century, gave Americans a sense of the way things could be—a cause for hope. The Labor Museum at Hull House, the Philadelphia Commercial Museum, and the Newark Museum, for example, all exhibited artisans at work in an effort to demonstrate the dignity of labor and the worth of the applied arts, while international expositions likewise featured farmers and blacksmiths at work amidst the plaster splendor of Beaux-Arts exhibition halls.[23]

The aesthetic was as prominent in photographs and illustrations as it was in venues for live display. Francis Benjamin Johnston's serene images of students laboring at Hampton and Tuskegee, Helen Clark Perry's airy pen-and-ink sketches of potters and glassblowers, and Lewis Hine's work portraits all afforded viewers the opportunity to watch competent craftsmen at work in unfamiliar or unusual settings.[24] The American Museum of Natural History accumulated hundreds of photographs of indigenous people engaged in traditional crafts, working in front of the blank sheets that Franz Boas and George Hunt unfurled behind them, circulating them to schoolchildren, and Edward Curtis produced hundreds of images of "primitive" painters, carvers, and potters, quietly absorbed in their work.[25] Slides and stereographs produced by the Keystone View Company, Underwood & Underwood, and the Detroit Publishing Company featured skilled makers of objects working in home studios, tool-strewn workshops, or more unusual locales, as did newspapers like the *Chicago Daily News* and the *New York Times*, and magazines ranging from *The Craftsman* to *The Survey*. By the 1930s, F.S.A. photographers were training their lenses on artisans deep in concentration, as were picture magazines like *Life* and *Look*, and photo studios like Harris & Ewing.

The craftsmanship aesthetic had a particularly firm hold on visual representations of business in the interwar United States.[26] Newspapers, magazines, and news services used images of artisans to animate illustrations of newly developed professions and corporate entities.[27] Corporate advertisements also began to make use of this aesthetic. To create an emotional bond with potential consumers, advertisements of the era commonly featured skilled workers whose tasks suggested tradition rather than the deadening repetition that characterized most assembly-line work. Institutional photographers deployed the craftsmanship aesthetic as they recorded the work of the individuals who made up the massive labor forces of corporations like Western Electric or Shelton Looms.[28] When recording workers with objects that were not inherently visually compelling, photographers made increasing use of modernist techniques to make common settings and tools less familiar, using geometry, repetition, and abstraction to force viewers to see in new ways.

The aesthetic possessed tremendous popular appeal. At museums and fairs, visitors made a beeline for displays featuring craftspeople at work, for, as Labor Museum curator Jessie Luther observed, "the spot which attracts most people at any exhibition, or fair, is the one where something is being done," and organizers could expect that such activities "will almost inevitably attract a crowd, who look on with absorbed interest."[29] The skill of the laborers, their ability to transform raw material into something beautiful or useful, seemed to enchant onlookers. And the chance to glimpse a normally inaccessible setting, or see workers pursuing their craft in an unexpected environment, functioned as another appealing element of this aesthetic. At Newark and at the Commercial Museum in Philadelphia, for instance, the incongruity of craft activity within a museum's still halls excited visitors. According to one journalist, middle-class visitors found Hull House's Labor Museum altogether exotic, enjoying the chance to stroll amidst the other half, while the immigrant poor found the "glorified workshops, which promised pleasantness and peace" equally mysterious.[30]

Showing Making at the American Museum

Early twentieth-century photographers of the American Museum of Natural History capitalized on the craftsmanship aesthetic as they chronicled the construction of the museum's iconic exhibits. After 1900, taxidermists, modelers, painters, and other exhibit makers at the AMNH created dazzling reproductions of habitats in order to interest the public in the natural world.[31] Glass flowers bloomed, sea birds wheeled, and wolves leapt towards their prey, exciting much visitor conjecture about how the museum had managed to reproduce nature so realistically.[32] Staff photographers churned out thousands of hand-colored photographs and slides of these exhibits, which were loaned or sold to school systems, churches, and community groups.[33] Such images were also used to drum up publicity. Museum administrators placed them in house publications and provided photo editors of local and national periodicals with copies of these photographs.[34]

Staff photographers began to chronicle exhibit making at the museum in the 1910s, tapping into the public's ongoing interest in the production of artful and deceptive objects. The museum's publicity brochures, exhibit guides, and house organ, the *American Museum Journal*, occasionally featured images of exhibit making alongside pictures of completed exhibits. Readers could look into the museum's studios to see hippos peeking out from dripping plaster and preparators summoning frogs from wax. Throughout the ensuing three decades, the museum's photographers regularly recorded the slow process of exhibit making, producing almost as many photographs of the process as they did of the finished results. Museum staff drew on these images to illustrate popular weekend lantern-slide lectures, and in the 1920s and 1930s, editors of national publications built illustrated features and photo essays around them. Featured in the *American Museum Journal*, rotogravure sections of newspapers like the *New York Times* and the *Washington Post*, and magazines ranging from *Life* to *Popular Mechanics*, these photographs of exhibit making depicted materials being transformed, magic being made (Figure 4).[35] The images were also featured in temporary exhibitions on exhibit making.

By providing glimpses of exhibits in progress, photographers entertained viewers while shoring up the museum's reputation.[36] Bombarded by new kinds of visual experience, early twentieth-century Americans tended, to borrow a phrase from Michael Leja, to "look askance" at show or spectacle—a skepticism that extended to the goods on display in institutions defined by their commitment to presenting scientific truths. Yet this pervasive doubt was accompanied by a fascination with the production of illusion. As a result, the museum's photographs of exhibit making simultaneously countered and reinforced visitors' gleeful suspicions. These images afforded viewers considerable creative agency, allowing them to examine, conjecture, and debate how exactly skin and bone became a lifelike mount. When arranged in series, the photos left less to the imagination,

Fig 4 "New Art Saves Strange Beasts: Pictures Tell Story of Preparing Exhibits for the Museum." *Popular Science Monthly*, vol. 120, January 1932, pp. 52–53. Courtesy Science, Industry & Business Library, The New York Public Library, Astor, Lenox and Tilden Foundations.

but still allowed viewers to appreciate the ingenious construction of uncannily natural effects.[37] These visual explications entertained viewers and played to continued interest in visual artifice as well as their self-consciousness about the act of looking.

But the images were not intended solely to entertain. By making visible the mechanics of its illusions, they confirmed the museum's commitment to truth and affirmed its devotion to public education. They conveyed an institutional determination to "bring to those who cannot explore or travel, who cannot go very far beyond their immediate environment, the whole world of nature."[38] The museum's president, Henry Fairfield Osborn, frequently proclaimed this ambition in person and print, but the photographs served as hard proof of the museum's commitment to this endeavor.

In the 1910s, the museum's photographers began to supplement the existing repertoire of objects in progress with portraits of individual craftsmen at work.[39] Photographers' turn to individual portraits may have resulted from explicit direction, for the publicity-savvy Osborn understood the power of personality: in the late 1910s and 1920s, he explicitly directed the museum's publicists to highlight the museum's "naturalist-explorers," some of whom were exhibit makers.[40] Photographers may also have begun to shoot portraits instinctively, believing they would appeal to colleagues or interest the public more than images that reduced exhibit makers to anonymous figures. By the 1930s and 1940s, such portraits had become staples of the museum's visual iconography.

Rather than depict exhibit makers as incidental or menial laborers in an impersonal factory setting, these images cast them as skilled craftsmen devoted to reproducing nature in loving detail. Deeply absorbed in the task at hand, they posed in middle distance in the center of the frame, surrounded by picturesque tools of the trade. These later images featured close-ups of preparators' faces and hands, implying that exhibit makers' most essential characteristics were intelligence, character, and skill, rather than the brute strength suggested by earlier full-body depictions. The photographers drew on traditional iconographies of craftsmanship, echoing the occupational portraiture that emerged in the late eighteenth and early nineteenth century. Just as these portraits had made a case for the importance of the skilled craftsman to the civic virtue of the young republic, images of exhibit makers celebrated the contributions of the skilled craftsman to the museum, and, by extension, to the civic project of public education. In these portraits, however, exhibit makers did not make the kind of eye contact with the viewer that artisans in earlier portraits did. Instead, they were presented as so dedicated that they never took their eyes off their work.

Though the photographs were formulaic, they weren't static. In the 1910s and early 1920s, photographers granted exhibit makers a specific artistic status, positioning them somewhere between mediums and magicians. In Thomas Lunt's 1910 pictures of sculptor Sigurd Neandross and modeler Mr. Shimatori, for instance, the absorbed preparators seem to see far beyond the surface of the objects they hold (Figure 5).[41] The dreamy focus and metaphorical qualities of these photographs were certainly influenced by the pictorialist aesthetic so popular with American photographers, and such images served to reinforce long-held Romantic ideas about the inherently communicative quality of objects from the natural world. They reflected the widespread belief among certain members of the museum staff that the best exhibit makers had an extraordinary ability to understand, even empathize with, specimens.[42] Exhibit makers were so familiar with specimens, suggested these photographs, that they could elicit not just the shape but also the spirit of a plant, animal, or human.

In the 1930s and after, photographers moved away from this romantic style, instead casting exhibit makers as accomplished technical experts. These photographs depicted preparators in spotless studios, seated at well-lit tables, surrounded by neatly organized and well-sharpened instruments.[43] Photographers traded the soft focus of pictorialism for the straight, sharp focus associated with modernist, documentary, and scientific photography. Lighting had been diffuse but was now concentrated, and, in later images, almost clinical in its intensity. In Charles Coles's 1939 image, preparator Armin Schmidt, surrounded by tools as he carefully inserts a wooden brace into the

Fig 5 Thomas Lunt, "Sigurd Neandross modeling figure for use in ceremonial canoe," 1910. Image #22009, American Museum of Natural History Library.

Fig 6 Charles H. Coles, "Mr. Armin Schmidt closes the skin of the bird over the tow body," 1939. Image #291066, American Museum of Natural History Library.

small hollow of a bird's abdomen, looks very much like a surgeon—a profession that likewise merged scientific knowledge, manual skill, and intuition (Figure 6).[44] Indeed, these late images frequently adopted the iconography the public associated with professional science and medicine. Rather than overalls or paint-splattered artist's smocks, exhibit makers now wore lab coats or spotless shirtsleeves and ties. Rather than picturesque studios filled with props and half-draped portrait screens, exhibit makers were now shown in more sterile spaces, cleansed of distractions by lighting and tight cropping. In the 1930s and after, photographers often granted viewers an elevated perspective while still allowing them to come close, placing the lens over the technicians' shoulders to help the viewer better appreciate their precision and skill.

Exhibit making was a classic example of what David Pye has described as the "workmanship of risk," in which the quality of the object produced depends entirely upon the skill, judgment, and care exercised by its maker. Reproducing nature was a tricky business.[45] A skin spoiled by imprudent packing or an errant knife could not be replaced. Too much breath, and a spiky glass radiolarian would be blown out of proportion. Too little subtlety, and a diorama background would look flat. The materials themselves provided considerable challenges for even the most careful exhibit maker. Hides shrank or discolored, flowers died and leaves dried up, sun and rain rotted carcasses before an expedition's end. Consequently, exhibit making required long experience, artistic skill, and ingenuity—the hallmark traits of a master craftsman. An exhibit maker had to identify what exactly it was that had made that elephant look alive, and then, relying upon his sculptural talents and his knowledge of pachyderms and the material properties of their body parts, conjure the elephant out of glass, plaster, scaffolding, and a huge blanket of elephant skin stiffened by salt and time.[46] Without highly skilled taxidermists

and preparators, it was impossible to produce such objects. Neither machines nor systematic methods could replace these makers.

Yet the images could only hint at the infinite complexity of the craft of exhibit making, dictated as it was by the shape and variety of nature itself. Exhibit makers were responsible for preserving or recreating anything: jellyfish, flamingos, granite, spruce needles, moose, lilacs, tree frogs.[47] They relied on an extraordinary range of tools to do so. A good library was indispensible; so too was a "good stomach and a clear head," according to Ohio museum taxidermist Oliver Davie, because the work could be disgusting and dangerous.[48] The American Museum sent taxidermists and artists all over the globe to collect specimens and impressions of their habitats, so shotguns, compasses, knives of all sizes, rulers, cameras, plates, tripods, sketchbooks, canvas, paint, and easels were part of an exhibit maker's toolkit. Taxidermists used poisons—arsenic, denatured alcohol, copper subacetate, chlorine, technical acetone—to preserve stinking hides and bones. They also relied on homelier substances, using corn meal and milk to draw fat from duck carcasses, turpentine to clean feathers, and chloroform to stun snakes into submission long enough to make plaster casts of their coils. Modelers depended upon wax, paint, glycerin, and glass to reproduce plant life or create magnified models of anything microscopic. Taxidermists also used these materials, plus clay and plaster, varnishes, wire, scrapers, chisels, drills, calipers, pliers, and needles to sculpt muscular animal bodies, over which prepared skins could be draped. Exhibit making was scientific, mechanical, architectural, and

artistic; but it remained, very distinctly, craft work.[49]

Exhibit makers' ingenuity and craftsmanship, combined with the inherent visual incongruities of their work—its tiny or grand scale, partial mimesis, and frequent juxtaposition of the mundane and the exotic—proved irresistible to both photographers and viewers. While amusing themselves with the minute or the magnificent, the studio or the safari, viewers could also soothe themselves with visions of trustworthy, knowledgeable, and highly skillful craftsmen at one with their work. Museum photographers made the most of the visual absurdities of exhibit-making, occasionally tempering preparators' earnestness by depicting them in circumstances that seem intended to provoke hilarity or amazement. In one, the elderly Charles Lang and Otto Falkenbach, two preparators for the museum's Department of Vertebrate Paleontology, serenely work on an enormous but equally wizened triceratops head—a witty visual meditation on the theme of old age (Figure 7).[50] In another, preparator Robert H. Rockwell, whose diminutive stature was incommensurate with his considerable talent, was made to seem even smaller as he stretched up to model the neck of a bull giraffe.[51] Museum photographers captured background painter Chris Oleson hard at work with canvas and brush, underwater, wearing diving gear.[52] But even in those images that encouraged joking or gawking at the peculiarities of museum work, photographers celebrated these artisans and never compromised their dignity.

Photographs of exhibit makers added luster to the AMNH's reputation. The museum's willingness to shelter craft for the

Fig 7 Julius Kirschner, "Charles Lang and Otto Falkenbach working on Triceratops model," 1938. Image #315711, American Museum of Natural History Library.

purposes of public education, as impractical as it seemed in the machine-minded twentieth century, held tremendous public appeal. Though the Arts and Crafts movement had collapsed by the First World War, its underlying search for a sense of wholeness through artistic labor endured, and critics lauded the museum's commitment to artful science. As Joseph Moncure March noted in a 1925 *New York Times* article, the museum's exhibits were made "with two hands and ten fingers—not forgetting patience and skill."[53] "This, in an age where there are telephones and subways and mass production and speed at all costs, gives one a strange and not altogether unrefreshing feeling," March mused.[54] Photographs of exhibit making that featured the craftsmanship aesthetic bolstered the idea that the museum—now humanized by visual depictions of its staff members—would go to any lengths to ensure its exhibits rendered nature in scrupulously accurate terms. The craftsmanship aesthetic in these photographs suggested that the AMNH's exhibits were absolutely believable, for they were no cheap artifices or machine-made copies, but individualized products of hours of honest labor, arduous self-sacrifice, and passion for craft.

The Uses of the Craftsmanship Aesthetic at the AMNH

For the most part, the craftsmanship aesthetic helped sustain the ascending reputation of exhibit makers within the museum itself. Nineteenth-century preparators, assigned the menial work of scraping, preserving, and "stuffing" animal skins, had been treated as glorified janitors by many of the museum's curators. Exhibit makers' status improved considerably as professional artists, modelers, and taxidermists joined their ranks, and the museum's photographs both reflected and encouraged this development. Compared "to the men who often bear the title of taxidermist," the current exhibition staff stood as "[August] Rodin and [Daniel Chester] French [stood] to the cave artists," declared the *American Museum Journal* in 1913, and photographers attempted to bear this comparison out.[55] Throughout the next two decades, images documenting exhibits in process gradually declared exhibit makers' technical ingenuity, artistic skill, and scientific knowledge.

Photographs of exhibit makers, occasionally featured in the museum's in-house periodicals as well as its exhibits and public relations, may have played a subtle role in institutional political conflicts. In the late 1920s and early 1930s, exhibit makers pressed for more institutional recognition of their worth and more independence from curatorial oversight. A decade-long power struggle between preparators and curators ensued.[56] Images of exhibit makers betray where photographers' sympathies lay in this internal battle. By proclaiming the artistic and intellectual expertise of the museum's

taxidermists, modelers, and painters, the photographs provide a visual argument for the ongoing validity of what Pamela Smith has described as "artisanal epistemology."[57] In a 1930 photograph captioned "Barnum Brown and Otto Falkenbach preparing new (unnamed) reptile," for instance, Brown, a curator, and Falkenbach, a preparator, seem to be engaged in a shared pursuit of knowledge. Engrossed in their work, both Brown and Falkenbach rely on lenses to examine the specimens in front of them (Figure 8). Falkenbach's magnifying glass, work coat, and battered desk evoke an older kind of scientific pursuit than Brown's shiny microscope, suit, and working surface, but the photograph grants their focused labor similar respect, even as it hints they might not be receiving equivalent financial compensation from the museum.

Yet images often serve multiple, even contradictory, purposes, and while the craftsmanship aesthetic trumpeted the virtues of the museum's artisans and their labor, it could be used to placate the same people it praised. Indeed, these triumphant photographs, commissioned by the museum's administration and published in house periodicals, may have provided administrators with at least one easy way of gratifying exhibit makers' egos without upending the museum's longstanding hierarchy of scientific authority.[58] The craftsmanship aesthetic could also be excised from reproductions of these carefully composed images. In the spate of articles on exhibit making that ran through the 1930s in *Popular Science*, *Popular Mechanics*, and *Boys' Life*, for instance, the fascinations of exhibit making trumped the importance of exhibit makers or the

Fig 8 H.S. Rice and Irving Dutcher, original caption: "Barnum Brown and Otto Falkenbach preparing new (unnamed) reptile," 1930. Image #313156, American Museum of Natural History Library.

appeal of their settings. Though the images in these publications were often drawn directly from the American Museum's photographic morgue, editors' interest in the technology and processes of production frequently overwhelmed their taste for the craftsmanship aesthetic, so they eliminated it through cropping, resizing, or collage. In 1937, *Life*'s photo editors selected four images from a 1927–8 series of photographs depicting the creation of a giraffe mount, and published them, comic strip style, across the top third of a single page. While the reproduced pictures nicely captured the technical process of creating a plaster

cast upon which the giraffe's skin could be draped, their small size reduced exhibit makers to visual props that indicated the giraffe's colossal height. The craftsmanship aesthetic that had characterized so many of the original images in the series was no longer discernible (Figure 9).

Nonetheless, museum photographers continued to incorporate the craftsmanship aesthetic into their images, and may have done so to increase respect for museum-based artists. Throughout the interwar decades, American artists of all types pushed corporations, civic institutions, and the broader public to recognize their

Fig 9 "Stuffing a Giraffe." *LIFE*, May 25, 1937. Images #312076, #312179, #312201, #314065, American Museum of Natural History Library.[59]

worth.[60] Some artists formed professional organizations; others attempted to unionize. Inside and outside the museum, as they struggled to win the respect they believed their profession deserved, artists sought to represent their work as dignified, difficult, and essential. While artists sympathetic to the labor movement often identified themselves as workers in these years, many of those inhabiting less established realms of artistic production—commercial art, industrial design, craft, and photography—aspired to the very status of fine artist that many painters and sculptors were attempting to escape.[61] The AMNH's exhibit makers and photographers certainly fell into the category of those who aspired to public recognition as artists rather than laborers, and by depicting their exhibit making peers in a respectful light, museum photographers may have believed they too would garner more respect for themselves and their craft.

In the 1930s and 1940s, the museum's ambitious young photographers increasingly employed both the aesthetics of craftsmanship and the technological sublime in self-depictions—presumably in an effort to improve their own institutional status. In one 1942 image, photographer Thane L. Bierwert portrays Charles H. Coles photographing glassblower Herman O. Mueller and one of Mueller's delicate glass models from above (Figure 10).[62] Mueller's white hair and pale hands and the spiny glass radiolarian, all lit by carefully positioned spotlights, glow brightly. But it is immediately clear this shot was not intended to showcase the model or to lionize the maker. In this image, Coles, not Mueller, is the one at work, and it is clear that he and Bierwert have staged the shot to celebrate their craft, their skilled labor. The tools of the photographic trade are prominently placed in the image: the three spotlights that surround Mueller, the accordion portrait camera, the ladders, the timer carefully angled toward the camera lens, and, most obviously, the equipment crate clearly labeled "Division of Photography" upon which Coles stood. The instruments signal that there is nothing casual about the work of museum photographers—that each shot was crafted carefully, as delicate a construction as the snowflake-like sphere resting in front of

Fig 10 Thane L. Bierwert, "Glass blower Herman O. Mueller and photographer Charles H. Coles, photographing radiolarian," 1942. Image #29692, American Museum of Natural History Library.

Mueller. Ostensibly an image of Mueller and model, the image is really a photograph about photographing.

It is also an image that implicitly reminds viewers of the importance of photography to the museum. Like the exhibit makers in Bierwert's photographs, Coles seems thoroughly engaged, cupping the camera lens, half-smiling as he bends forward and beneficently gazes down on his subject. He is elevated, reminding the viewer of the power inherent in his ability to represent—and beatify—the museum's work. Coles's posture mirrors that of the eye-like spotlight on the

other side of the photograph, suggesting his vision, no less than the light itself, is necessary to illuminate the work of the American Museum for the public. However quietly self-aggrandizing the image might have been, there was truth in it. In the late 1930s and 1940s, the Depression and war had drained the museum's once bursting coffers, and inept administrators and constant infighting had demoralized the staff.[63] Photography helped keep the public face of the museum unblemished, even glowing, and this photograph serves as a pointed reminder of that fact.

The Journal of Modern Craft   Volume 5—Issue 1—March 2012, pp. 25–50

Why Bierwert and Coles took this particular image remains a mystery. It may have been an attempt to persuade administrators to provide more funding, or perhaps it was used as an illustration in the photo staff's periodic lectures on camera work to their colleagues at the museum.[64] It may have been a private indulgence, a still staged during a photo shoot of Mueller, whose work was frequently featured in illustrated periodicals. The museum's photographers, who understood the formidable power of its photo archive, may not have had a specific contemporary viewer in mind when they took it. Perhaps they hoped to remind future staff members, scholars, even the thousands of schoolchildren subjected to the museum's decades-old slide collections, of the largely invisible but entirely necessary craft of museum photography. Or perhaps it was intended as what is known in painting as an "easel image," a picture meant to be appreciated in the original by a select few, rather than seen via mass reproduction in magazines or on slides. This seems quite possible: makers have always been conscious of the way they are depicted. Centuries of artists' self-portraits testify to this, as do the requests made of artists in the business of showing making. When blacksmith Pat Lyon commissioned John Neagle to paint his portrait, he did so with specific instructions, and his request to be painted "at work, at my anvil, with my sleeves rolled up and a leather apron on" indicated both his own beliefs about the importance of craft labor, and a perceptive grasp of the American public's belief in the worth of craft and the incorruptibility of craftsmen.[65]

One century and two industrial revolutions after Lyon made his request, Americans felt even more strongly that craft and craftsmanship, regardless of where it was found, retained both visual appeal and moral value. Yet these photographs and the aesthetic they capitalized upon may have given exhibit makers a sense that they were at least seen, if not heard. They also offered a form of empty therapy, satisfying Americans with images or performances of craftsmanship without providing them with its personal rewards. Images could not replace the pecuniary benefits in which craft was once a viable economic option, nor could it fully provide the emotional or personal satisfaction that so many Americans seemed to find in making things. But the craftsmanship aesthetic was resonant nonetheless—and has remained so throughout the twentieth century, as is made evident by the success of popular figures like Julia Child, Bob Ross, and Bob Vila, not to mention the Home and Garden channel, the Food Network, and PBS. Corporations continue to exploit the craftsmanship aesthetic as a way to hawk goods, even to the point of satirizing it—in a 2010 advertisement for Velveeta Cheesy Skillets, for instance, the pride and conviction of a nineteenth-century blacksmith intent on creating an excellent meal bewilders a mother merely trying to get dinner on the table that night.[66] In the museum context, iterations of this aesthetic have remained visual staples of the American Museum of Natural History's public relations efforts and institutional record, and its photographers continue to draw upon this aesthetic to this very day.

Acknowledgments

I am indebted to the Mellon Foundation, Vanessa Schwartz, and the University of Southern California, whose support provided me with the time and resources to conduct the research for the conference paper that served as the basis for this article, and to Eric Foner, Irene Cheng, Sarah Goodrum, Sarah Miller, and Seth Weingram, whose comments and suggestions helped shape this piece. Great thanks to Ann-Sophie Lehmann for inviting me to contribute to the conference and this issue, and to Glenn Adamson, Ned Cooke, and a skillful and generous anonymous peer reviewer for providing feedback that significantly improved this article. Thanks as well to Barbara Mathe, for her expertise and supervision of the photographic archive of the American Museum of Natural History, and to Alexandra Cain, whose work inspired this article.

Notes

1 As cited in Nicolas Maffei, "John Cotton Dana and the Politics of Exhibiting Industrial Art in the United States, 1909–1929," *Journal of Design History*, 13(4) (2000): 308–9.

2 On the Newark Museum's industrial and craft exhibits and the institution's broader Progressive efforts to educate workers, see Carol Duncan, *A Matter of Class: John Cotton Dana, Progressive Reform and the Newark Museum* (Pittsburgh, PA: Periscope Publishing, 2009), pp. 113–39; Maffei, "John Cotton Dana and the Politics of Exhibiting Industrial Art in the United States, 1909–1929."

3 On the social impact of large industrial corporations on American workers, see, among others, Davin Noble, *America by Design: Science, Technology, and the Rise of Corporate Capitalism* (New York: Oxford University Press, 1979); David Montgomery, *The Fall of the House of Labor: The Workplace, the State, and American Labor Activism, 1865–1925* (New York: Cambridge University Press, 1987). On the diversity of production methods and environments in the second industrial revolution, see Philip Scranton, *Endless Novelty: Specialty Production and American Industrialization, 1865–1925* (Princeton, NJ: Princeton University Press, 2000). On the broad cultural effects of Fordism, Taylorism, and corporate interest in efficiency, see Lewis Mumford, *Technics and Civilization* (New York: Harcourt Brace & Company, 1934); Anson Rabinbach, *The Human Machine: Energy, Fatigue, and the Origins of Modernity* (New York: Basic Books, 1990); Martha Banta, *Taylored Lives: Narrative Productions in the Age of Taylor, Veblen, and Ford* (Berkeley and Los Angeles, CA: University of California Press, 1993).

4 The literature on this anxiety is enormous, and includes James B. Gilbert, *Work without Salvation: America's Intellectuals and Industrial Alienation, 1880–1910* (Baltimore, MD: Johns Hopkins University Press, 1977); David Brody, "The American Worker in the Progressive Age," in David Brody (ed.), *Workers in Industrial America* (New York: Oxford University Press, 1993); Herbert Gutman, *Work, Culture, and Society in Industrializing America* (New York: Alfred A. Knopf, 1976). Attention to the impact of this trend on visual culture has also garnered considerable scholarly interest. Eileen Boris and Jackson Lears have ably chronicled how the American arts and crafts movement sought to respond to this development, while Terri Smith, Alan Trachtenberg, Elspeth Brown, and Melissa Dabakis have explored the ways that American visual culture absorbed and articulated fears about modern industrial labor. Eileen Boris, *Art and Labor: Ruskin, Morris, and the Craftsman Ideal in America* (Philadelphia, PA: Temple University Press, 1986); T.J. Jackson Lears, *No Place of Grace: Antimodernism and the Transformation of American Culture, 1880–1920* (Chicago, IL: University of Chicago Press, 1981); Terri Smith, *Making the Modern: Industry, Art, and Design in America* (Chicago, IL: University of Chicago Press, 1993); Alan Trachtenberg, *Reading American Photographs:*

Images as History, Mathew Brady to Walker Evans (New York: Hill and Wang, 1990); Elspeth Brown, *The Corporate Eye: Photography and the Rationalization of American Commercial Culture* (Baltimore, MD: Johns Hopkins University Press, 2005); Melissa Dabakis, *Visualizing Labor in American Sculpture: Monuments, Manliness, and the Work Ethic, 1880–1935* (New York: Cambridge University Press, 1999).

5 On the development of the consumer culture of the United States in the early twentieth century, see, among others, T.J. Jackson Lears, *Fables of Abundance: A Cultural History of Advertising in America* (New York: Basic Books, 1995); William Leach, *Land of Desire: Merchants, Power, and the Rise of a New American Culture* (New York: Vintage Books, 1993).

6 On the ongoing efforts of consumer institutions to efface understanding or shape public perceptions of production, see, for instance, Roland Marchand, *Creating the Corporate Soul: The Rise of Public Relations and Corporate Imagery in American Big Business* (Berkeley and Los Angeles, CA: University of California Press, 1998).

7 Michael Leja, *Looking Askance: Skepticism and American Art from Eakins to Duchamp* (Berkeley, CA: University of California Press, 2004), p. 1.

8 David Nye, *American Technological Sublime* (Cambridge, MA: The MIT Press, 1994), pp. 127–42.

9 As Richard Sennett did in his recent sociological study of craftsmanship, I define the concept of the craftsman and craftsmanship broadly, defining it as an approach to labor—the attitude, skill, and experience it takes "to do a job well for its own sake," to use Sennett's phrase—rather than as a particular constellation of activities. See Richard Sennett, *The Craftsman* (New Haven, CT: Yale University Press, 2008), p. 9.

10 Neil Harris, *Humbug: The Art of P.T. Barnum*, Phoenix ed. (Chicago, IL: University of Chicago Press, 1981), p. 57. A more extensive discussion of the operational aesthetic can be found on pages 59–91.11 On American responses to technology in the mid-nineteenth century, see, among others, Nye, *American Technological Sublime*. On contemporary preoccupations with visual deception, see Karen Haltunnen, *Confidence Men and Painted Women: A Study of Middle-class Culture in America, 1830–1870* (New Haven, CT: Yale University Press, 1986); Miles Orvell, *The Real Thing: Imitation and Authenticity in American Culture* (Chapel Hill, NC: University of North Carolina Press, 1989); James W. Cook, *The Arts of Deception : Playing with Fraud in the Age of Barnum* (Cambridge, MA: Harvard University Press, 2001).

12 The introduction of electricity into the production of goods and the scale of industrial and mass production likewise precluded easy comprehension. Nye, *American Technological Sublime*, p. 126.

13 On the specialization of knowledge in the late nineteenth and early twentieth century United States, see Alexandra Oleson and John Voss (eds), *The Organization of Knowledge in Modern America* (Baltimore, MD: The Johns Hopkins University Press, 1979); Robert H. Wiebe, *The Search for Order 1877–1920* (New York: Hill and Wang, 1967).

14 Harris, *Humbug: The Art of P.T. Barnum*, p. 74; Henry Adams, *The Education of Henry Adams: An Autobiography* (Boston, MA: Houghton Mifflin, 1973; original work published in 1918), p. 374.

15 Robert S. Lynd and Helen Merrell Lynd, *Middletown: A Study in Modern American Culture* (New York: Harcourt Brace and Company, 1929), p. 248.

16 Ibid., p. 271.

17 Ibid., p. 154.

18 Harry R. Rubenstein, "With Hammer in Hand: Working Class Occupational Portraits," in Howard B. Rocke, Paul A. Gilge, and Robert Asher (eds), *American Artisans: Crafting Social Identity, 1750–1850* (Baltimore, MD: Johns Hopkins University Press, 1995), pp. 178–98.

19 As cited in Bruce W. Chambers, "The Pythagorean Puzzle of Patrick Lyon," *The Art*

Bulletin, 58(2) (1976): 225. Lyon's portrait of the artisan at work in his forge was painted by John Neagle in 1826–7 and is now in the Museum of Fine Arts, Boston.

20 On images of craftsmen in the United States, see Laura Rigal, *The American Manufactory: Art, Labor, and the World of Things in the Early Republic* (Princeton, NJ: Princeton University Press, 2001); Wayne Craven, *Colonial American Portraiture* (New York: Cambridge University Press, 1986); H. Barbara Weinberg and Carrie Rebora Barratt (eds), *American Stories: Paintings of Everyday Life, 1765–1915* (New York: Metropolitan Museum of Art, 2009).

21 On anti-modernist fixations on the artisan, see Lears, *No Place of Grace: Antimodernism and the Transformation of American Culture, 1880–1920*, pp. 60–96. Both Lears and Eileen Boris point out that much of this contemporary interest in artisanal labor was delusional, as few artisans were able to eke out even a subsistence living from their crafts. Boris, *Art and Labor: Ruskin, Morris, and the Craftsman Ideal in America.*

22 On craft demonstrations on Indian reservations, see Elizabeth Hutchinson, *The Indian Craze: Primitivism, Modernism, and Transculturation in American Art, 1890–1915*, ed. Nicholas Thomas, Objects/Histories (Durham, NC: Duke University Press, 2009). On World's Fairs craft demonstrations, see, among others, Matthew F. Bokovoy, *The San Diego World's Fairs and Southwestern Memory, 1880–1940* (Albuquerque, NM: University of New Mexico Press, 2005), p. 115. Deepali Dewan has noted a similar phenomenon in Britain and colonial India in this period: see Deepali Dewan, "The Body at Work: Colonial Art Education and the Figure of the 'Native Craftsman,'" in James H. Mills and Satadru Sen (eds), *Confronting the Body: The Politics of Physicality in Colonial and Post-Colonial India* (London: Anthem Press, 2004).

23 On live craft demonstrations at the Labor Museum, see Marion Foster Washburne, "A Labor Museum," *The Craftsman*, 6 (1904); Jessie Luther, "The Labor Museum at Hull House," *The Commons*, 70(7) (1902). On the Commercial

Museum of Philadelphia, see Stephen Conn, *Museums and American Intellectual Life, 1876–1926* (Chicago, IL: University of Chicago Press, 1998). On "process demonstrations" at the Newark Museum, see Carol Duncan, *A Matter of Class: John Cotton Dana, Progressive Reform, and the Newark Museum* (Pittsburgh, PA: Periscope Publishing, 2009), p. 115; Ezra Shales, *Made in Newark: Cultivating Industrial Arts and Civic Identity in the Progressive Era* (New Brunswick, NJ: Rutgers University Press, 2010); Maffei, "John Cotton Dana and the Politics of Exhibiting Industrial Art in the United States, 1909–1929," pp. 308–9.

24 On Johnston's photographs of Hampton and Tuskegee, see Bettina Berch, *The Woman behind the Lens: The Life and Work of Frances Benjamin Johnston 1864–1952* (Charlottesville, VA: University Press of Virginia, 2000), pp. 33–69. On Perry, see Duncan, *A Matter of Class: John Cotton Dana, Progressive Reform, and the Newark Museum*, pp. 121–5. On Hine and his work portraits, see Alan Trachtenberg and Walter Rosenblum, *America and Lewis Hine* (New York: Aperture, 1984); Elspeth H. Brown, *The Corporate Eye: Photography and the Rationalization of American Commercial Culture* (Baltimore, MD: Johns Hopkins University Press, 2005), pp. 119–58; Kate Sampsell-Willmann, *Lewis Hine as Social Critic* (Jackson, MS: University Press of Mississippi, 2009).

25 Aaron Glass, "Frozen Poses: Hamat'sa Dioramas, Recursive Representation, and the Making of a Kwakwaka'wakw Icon," in Elizabeth Edwards and Christopher Morton (eds), *Photography, Anthropology and History: Expanding the Frame* (Burlington, VT: Ashgate, 2009); Alison Griffiths, *Wondrous Difference: Cinema, Anthropology and Turn-of-the-century Visual Culture*, ed. John Belton, Film and Culture (New York: Columbia University Press, 2002); Ira Jacknis, "Boas and Photography," *Studies in Visual Communication*, 10 (1984); Edward S. Curtis, *The North American Indian* (London: Taschen, 2003).

26 In the 1930s, for instance, Harvard Business School professor Donald Davenport and Frank Ayres, Executive Secretary of the Business

Historical Society, compiled more than 2,000 publicity images from 115 different corporations to illustrate "the human factor" in American manufacturing for their students, and both the craftsmanship and the operational aesthetic are evident throughout this collection. See Melissa Banta, "The Human Factor: Introducing the Industrial Life Photograph Collection at the Baker Library" (Harvard Business School, 2010).

27 Elspeth Brown's excellent discussion of this phenomenon and her associated discussion of "capitalist realism" has been indispensible to the development of my own argument. See Brown, *The Corporate Eye: Photography and the Rationalization of American Commercial Culture*, especially pp. 119–58; Marchand, *Creating the Corporate Soul: The Rise of Public Relations and Corporate Imagery in American Big Business*; David E. Nye, *Image Worlds: Corporate Identities at General Electric, 1890–1930* (Cambridge, MA: The MIT Press, 1985).

28 See Banta, "The Human Factor: Introducing the Industrial Life Photograph Collection at the Baker Library"; Lewis Hine, *Men at Work: Photographic Studies of Modern Men and Machines* (New York: Macmillan Company, 1932).

29 As cited in Washburne, "A Labor Museum," p. 573.

30 Ibid., p. 570.

31 On the history of habitat dioramas in American natural history museums, see, among others, Karen Wonders, *Habitat Dioramas*, Figura Nova Series 25 (Acta Universitatis Uppsaliensis, 1993); Hanna Rose Shell, "Introduction: 'The Soul in the Skin,'" in *The Extermination of the American Bison* (Washington, DC: Smithsonian Institution Press, 2002); Mary Anne Andrei, "Nature's Mirror: How the Taxidermists of Ward's Natural Science Establishment Reshaped the American Natural History Museum and Founded the Wildlife Conservation Movement" (dissertation, University of Minnesota, 2006). On the history of glass flowers and models, see Soraya de Chadarevian and Nick Hopwood (eds), *Models: The Third Dimension of Science* (Stanford, CA: Stanford University Press, 2004) and Lorraine Daston (ed.), *Things that Talk: Object Lessons from Art and Science* (New York: Zone Books, 2004).

32 Newspaper reports and the first two decades of papers given at the annual meeting of the Association of American Museums are filled with descriptions of such visitor conversation. See, for instance, Delores Bacon, "Where Art Rivals Nature," *New York Times*, November 10, 1901.

33 The first photographs of exhibits and exhibit making at the American Museum date from the early 1900s, and their relative scarcity and wide-ranging quality makes plain the ad hoc status of photography in the museum at that time. By the 1910s, however, the museum's photographic record had become less extemporaneous, as administrators hired professional photographers to chronicle the institution's activities. Hired under the auspices of the museum's Department of Public Education, which was charged with producing the thousands of visual aids the museum circulated each year, Thomas Lunt, Julius Kirschner, Kay C. Lenksjold, and their successors in the 1920s, 1930s, and 1940s—Charles H. Coles, Thane L. Bierwert, Lee Boltin, and others—served as the museum's photographers-at-large.

34 The *New York Times* and other New York City daily papers were ready repositories for images of the museum's exhibits and the expeditions launched to collect these images. In the 1900s and 1910s, illustrated magazines like *Century, World's Work,* and *Scribners'* occasionally published images of the American Museum's exhibits, but by the 1910s and 1920s, *Popular Science, Popular Science Monthly, Popular Mechanics* and *Boys' Life* published such images—either noting them as exhibits or using them to depict other scientific phenomena—on an almost monthly basis. *Time Magazine*, and later on, *Life*, proved equally eager to showcase the exhibits at the museum. In 1939 alone, *Life* featured the museum's exhibits in six different photo spreads.

35 Such photographs seemed to arouse enduring interest among magazine editors and presumably their readers: the work of American Museum glassblower Herman O. Mueller alone was the subject of several photographic features in picture magazines throughout the 1930s and 1940s, for instance. See, among others, "Glass Models of Ocean Life Fashioned by Blowing," *Popular Mechanics*, 58(2) (August 1932): 228; "Model Aquatic Life," *Life*, December 21, 1936, pp. 44–5; "Glass Sculptor Models Microscopic Life," *Popular Science*, 139(4) (October 1941): 121–2; "Modeling Marine Life," *Popular Mechanics*, 83(1) (January 1945): 72–3.

36 Throughout the 1910s and 1920s, corporate-sponsored industrial films on the production of goods, for instance, also sought to entertain audiences while educating them about the industry's importance and worth, or generating excitement about new products. Beginning in the 1930s, Hollywood studios adopted a similar tack, regularly producing "behind-the-scenes" and "making-of" featurettes as a sophisticated form of public relations. See Vinzenz Hediger, "Making Movies Is Like Making Cars, Only More Fun," in *Hollywood: Recent Developments*, ed. Christian W. Thomsen and Angela Krewani (London and Stuttgart: Edition Axel Menges, 2005).

37 Tom Gunning has referred to a similar phenomenon in the cinematic culture of the period, which he describes as part of an "aesthetic of astonishment." See Tom Gunning, "An Aesthetic of Astonishment: Early Film and the Incredulous Spectator," *Art & Text* (Spring 1989).

38 Henry Fairfield Osborn, "The American Museum and the World," *Fifty-Fifth Annual Report*, 1924, p. 1.

39 Vinzenz Hediger, writing about "making of" shorts sponsored by American movie studios, and Elspeth Brown, exploring employee magazines produced by Western Electric and other corporations, have also noted the emergence of individuals alongside technology and workspaces in corporate-sponsored publicity in the 1910s and early 1920s. See Hediger, "Making Movies Is Like Making Cars, Only More Fun," in *Hollywood: Recent Developments*, ed. Thomsen and Krewani, p. 57; Brown, *The Corporate Eye: Photography and the Rationalization of American Commercial Culture*, pp. 136–58.

40 John Michael Kennedy, "Philanthropy and Science in New York City: The American Museum of Natural History, 1868–1968" (dissertation, Yale University, 1968), p. 157.

41 Thomas Lunt, "Sigurd Neandross modeling figure for use in ceremonial canoe," November 1910 (Image 33009, AMNH Negative Logbook 17, AMNH-PC) and "Mr. Shimatori working on a model in the laboratory of the Department of Preparation and Installation," December 1910 (Image 33028, AMNH Negative Logbook 17, AMNH-PC).

42 On the history of the scientific knowledge of illustrators and artisans, see, among others, Lorraine Daston and Peter Galison, *Objectivity* (New York: Zone Books, 2007); Pamela H. Smith, *The Body of the Artisan: Art and Experience in the Scientific Revolution* (Chicago, IL: University of Chicago Press, 2004). On early twentieth-century ideas about the knowledge of exhibit makers in natural history museums, see Victoria Cain, "The Art of Authority: Exhibits, Exhibit-Makers, and the Contest for Status at the American Museum of Natural History, 1890–1940," *Science in Context*, 24(2) (2011).

43 In these later photographs, exhibit makers exemplify the category of science expertise that Lorraine Daston and Peter Galison have described as "trained judgment," a kind of instinctual, highly competent capability to visualize and render based upon long and repeated exposure to certain kinds of data. See Daston and Galison, *Objectivity*, pp. 309–61.

44 Charles H. Coles, "Armin Schmidt inserting tow body into skin for mounting bird," February 1939 (Image 291065, AMNH Negative Logbook 10, AMNH-PC) and "Mr. Armin Schmidt closing skin over the tow body for mounting bird" (Image 291066, AMNH Negative Logbook 10, AMNH-PC).

45 David Pye, *The Nature and Art of Workmanship* (New York: Cambridge University Press, 1968), p. 4. Exhibit makers' other most important attribute was creativity, for when they did not have the tools necessary for the job, they had to develop new ones. For instance, Carl Akeley, the museum's most famous taxidermist and an inveterate tinkerer, built a moving picture camera that kept its eyepiece level regardless of the camera's tilt, allowing him to record animal behavior even in the most challenging topographies. John Rowley, the museum's chief of exhibition in the 1890s and 1900s, found that a pan of paraffin and an old hairbrush could be used to create convincing drifts of snow, and used this knowledge to invent a machine that generated six inches of snow a minute. Brian Coe, *The History of Movie Photography* (Bernardsville, NJ: Eastview Editions, 1981), p. 82. Bacon, "Where Art Rivals Nature," SM10.

46 While women collected specimens, made models, and painted backgrounds alongside male exhibition staff, the museum's taxidermists were all male, probably as a result of the nature of the work—it was filthy, physically arduous, and frequently dangerous.

47 I have drawn heavily upon Susan Leigh Star's excellent overview of the craft of taxidermy in this paragraph. See Susan Leigh Star, "Craft vs. Commodity, Mess Vs. Transcendence: How the Right Tool Became the Wrong One in the Case of Taxidermy and Natural History," in Adele E. Clarke and Joan H. Fujimura (eds), *The Right Tools for the Job: At Work in Twentieth-century Life Sciences* (Princeton, NJ: Princeton University Press, 1992), pp. 257–86.

48 Oliver Davie, *Methods in the Art of Taxidermy* (Philadelphia, PA: David McKay, 1894), p. 16.

49 On the historically hybrid nature of craftwork in the modern era, see David McGee, "From Craftsmanship to Draftsmanship: Naval Architecture and the Three Traditions of Early Modern Design," *Technology and Culture*, 40(2) (1999): 209–36.

50 Julius Kirschner, "Charles Lang and Otto Falkenbach working on Triceratops model,"

August 1938 (Image 315711, AMNH Negative Logbook 19, AMNH-PC).

51 H.S. Rice and Irving Dutcher, February 1928, "Mr. Rockwell modelling the large male giraffe for the Waterhole Group, African Hall" (Image 312180, AMNH Negative Logbook 18, AMNH-PC). A series of images of Rockwell modeling this giraffe were published in *Life* magazine nine years later (see Figure 9).

52 See photographs printed with Malcolm L.M. Vaughan's article, "Painting Beauty Under the Sea," *The Los Angeles Times*, July 8, 1928, K1.

53 Joseph Moncure March, "Natural History That Rivals Nature—Wild Life in Marvelous Detail Is Shown in the Animal Groups at American Museum," The *New York Times*, February 1, 1925, SM15.

54 Ibid.

55 "The Work of Carl E. Akeley," *American Museum Journal*, 13(4/1913): 178. Artists had long attempted to assert—or at least share—authority over images they created for and with scientists. This ongoing conflict was almost certainly intensified by the professionalization of the art world in the first three decades of the twentieth century, as well as by exhibit makers' own independent artistic success. On artistic ambitions regarding scientific images, see Daston and Galison, *Objectivity*, pp. 84–98. On artistic ambitions among exhibit makers at the American Museum, see Cain, "The Art of Authority: Exhibits, Exhibit-makers, and the Contest for Status at the American Museum of Natural History, 1890–1940"; Michele Bogart, "Lowbrow/Highbrow: Charles R. Knight, Art Work and the Spectacle of Prehistoric Life," in T.J. Jackson Lears (ed.), *American Victorians and Virgin Nature* (Boston, MA: Isabella Stewart Gardner Museum, 2002).

56 Curators, already resentful of the resources devoted to exhibition and education rather than scientific research, accurately interpreted these demands as threats to their own hard-earned scientific authority. They fought to retain their position at the top of the museum's pecking order. See Cain, "The Art of Authority: Exhibits, Exhibit-makers, and the Contest for Status at the American Museum of Natural History, 1890–1940."

57 Smith, *The Body of the Artisan: Art and Experience in the Scientific Revolution.*

58 In her discussion of the use of photography by employee magazines in major corporations of the period, Elspeth Brown notes similar practices. See Brown, *The Corporate Eye: Photography and the Rationalization of American Commercial Culture*, pp. 148–9.

59 The original caption to the photo strip in *Life* read: "Stuffing a giraffe is no longer a matter of filling a skin with straw or wadding, setting it up in gawky, unconvincing lifelessness. Carl Akeley made taxidermy a sculptor's job. Above, an African Hall giraffe is prepared for an exhibit. At left, the sculptor builds a framework over which he fashions a soft modelling clay statue. From this, a plaster mold is made. The third picture shows the statues half torn away. The white edge is unremoved plaster mold. Inside the mold, layers of *papier maché* are carefully built up. When the mold is removed, the *papier maché* forms a hard, durable replica of the clay statue. Over this the giraffe skin is snugly fitted." © Time Inc. Used with permission. All Rights Reserved.

60 On commercial artists' struggle to be taken seriously in this period, see, among others, Michele Bogart, *Artists, Advertising and the Borders of Art* (Chicago, IL: University of Chicago Press, 1995).

61 On artists' sympathy with and self-identification as workers, see, among others, Michael Denning,

The Cultural Front (New York: Verso, 1997). On the struggle by designers, craftspeople, and photographers to attain professional status as artists, see Arthur Pulos, *American Design Ethic: A History of Industrial Design* (Cambridge, MA: The MIT Press, 1986); Joel Eisinger, *Trace and Transformation: American Criticism of Photography in the Modernist Period* (Albuquerque, NM: University of New Mexico Press, 1995); Bonnie Yochelson and Kathleen A. Erwin, *Pictorialism into Modernism* (New York: Rizzoli, 1996); Ellen Mazur Thompson, *The Origins of Graphic Design in America, 1870–1920* (New Haven, CT: Yale University Press, 1997).

62 Thane L. Bierwert, "Shooting radiolarian— Herman O. Mueller, Glass blower, Charles H. Coles, Photographer," October 1942 (Image 296922, AMNH Negative Logbook 10, AMNH-PC).

63 Kennedy, "Philanthropy and Science in New York City: The American Museum of Natural History, 1868–1968"; Clark Wissler, "Survey of the American Museum of Natural History: Made at the Request of the Management Board" (New York: The American Museum of Natural History, 1943).

64 Louis A. Monaco, Memorandum to the American Museum Employees' Camera Club, May 18, 1939 (Central Archives Box 1130.6-2, AMNH Archives).

65 Chambers, "The Pythagorean Puzzle of Patrick Lyon," p. 225.

66 See http://www.youtube.com/watch?v=ANIjAR-owjA&NR=1 (accessed September 13, 2011).

The Journal of Modern Craft

Volume 5—Issue 1
March 2012
pp. 51–68

DOI:
10.2752/174967812X13287914145479

Sacred Singularities: Crafting Royal Images in Present-day Thailand

Irene Stengs

Irene Stengs, a cultural anthropologist, is a Senior Research Fellow at the Meertens Instituut in Amsterdam, where she works on festive culture and ritual in the Netherlands. New public rituals, dynamics in lifecycle celebrations, and the relation between celebrity culture and everyday life in the Netherlands form her principal interests. She has also conducted research on modern cults, material culture, and social imaginary in urban Thai society. In 2009, Stengs published *Worshipping the Great Modernizer: King Chulalongkorn, Patron Saint of the Thai Middle Class*. She recently edited a book on the Dutchness of multicultural ritual in the Netherlands (*Nieuw in Nederland. Feesten en rituelen in verandering* 2012).

Abstract

The objective of this article is to address the significance of craftsmanship in the production and decoration of royal images in present-day Thailand, especially portraits of King Chulalongkorn (r. 1868–1910). In the 1990s, King Chulalongkorn became the object of a nationwide personality cult and as a consequence portraits of the king flooded the country. The explosion of portraits made worshippers want to create a distinction between their own objects and other reproductions, and, by implication, between themselves and other worshippers. Using Igor Kopytoff's notion of "singularization," a paradox is explained: in spite of virtually all portraits being mass-produced commodities, the owners of such images strive for their uniqueness. It is argued that crafting is a specific ritualized practice that singularizes the end product as a fine piece of craftsmanship. A vital aspect of this singularity is the object's making in the open, in shops or at markets, a practice that visually authenticates each single portrait produced.

Keywords: processes of authentication, Buddhist king(ship), craftmanship, Northern Thailand, Royal portraits, sacralization, singularization, woodcarving.

Bangkok, January 2008. At the entrance of Ratchadamnoenklang Avenue, a triumphal arch testifies to the ongoing celebrations for King Bhumibol Adulyadej (Figure 1). The arch is one of the thirty-one erected two years previously by the Bangkok Metropolitan Administration as part of the festivities for the king's sixtieth anniversary on the throne, an event that evolved seamlessly into the celebrations for the king's eightieth birthday on December 5, 2007.

The triumphal arch—a construction about ten meters high—consists of two separate archways, each spanning a lane of the wide avenue. The arches rest on a four-poled pavilion over the sidewalks. At first impression, the style is traditional Thai. Each arch carries seven golden crowns, and is festooned with gold and jewels. The decoration on top of the center pavilion

Fig 1 Triumphal arch in honor of King Bhumibol Adulyadej, Ratchadamnoenklang Avenue, Bangkok, January 2008. Photograph: Irene Stengs.

The Journal of Modern Craft   Volume 5—Issue 1—March 2012, pp. 51–68

consists of one of the royal Thai thrones flanked by two seven-tiered umbrellas, all in silver. Other decorations include guardian angels, mythical beasts, and two golden hares symbolizing 2007, the year of the rabbit. Most eye-catching, however, are over forty portraits of the king, together constituting the iconography of King Bhumibol as a meritorious and modern Buddhist king. Their octagonal frames glitter in the sun, as if set with huge diamonds. The realism of the full-color portraits contrasts with the mythical symbolism of the ornament. Yet, the most overwhelming effect made by these Thai religious and royal decorations—though they are traditionally crafted and covered in sumptuous gilding and mosaics of myriads of tiny, colored mirrors—is established by a very contemporary material. The shine and glitter are produced by colored CDs, probably recycled and then cut into rings and smaller discs.

In Thailand, images of King Bhumibol Adulyadej, or Rama IX, are omnipresent. Huge photographs of the king decorate government buildings, crossroads, bridges, verges, schools, and squares. Everywhere, in the cities and along provincial roads, in the mountains and in the fields, there are billboards, banners, flags, and posters carrying the royal image. In semi-public spaces such as shops, temples, offices, and restaurants, images of the king are on display, as in most people's private houses. The king's image is for sale on mundane objects as well: moneyboxes, clocks, jigsaws, calendars, buttons, and stickers, together constituting another public presence.

Most of the portraits of King Bhumibol are very similar. A small selection of images is used over and over again, and similarity can also be observed in their framings and decoration. Yet, amid this overwhelming homogeneity, some presences stand out— the triumphal arch, for instance. Apart from its size, the arch derives its outstanding appearance mainly from its craftsmanship. Nor is the arch an isolated phenomenon. Thai craftsmanship was traditionally connected with the court, nobility, and religious arts (cf. Apinan 1992; Peleggi 2002: 21–39). Today, the old crafts retain these associations but have also become part of popular middle-class culture. For the Thai urban middle class, buying or owning craft products forms a means to connect with cultural identity and establish class-specific distinctions.

The objective of this article is to address the significance of craftsmanship in the production and decoration of royal images in present-day Thailand, in the full awareness that this is but one dimension of a complex religious and political phenomenon. A central concept of my analysis will be Igor Kopytoff's notion of "singularization," the process by which a commodity's exchangeability is removed by rendering it unique, through the addition of particular values (Kopytoff 1986). Kopytoff's theory helps to resolve a seeming paradox: virtually all royal images are mass-produced commodities, but the owners of such images strive for uniqueness. The material for this investigation was collected during my research in the 1990s, which focused on the cult surrounding the grandfather of the present king, King Chulalongkorn (r. 1868– 1910), or King Rama V, modernizing ruler of what was then Siam.[1] Since the typical range of portraits of King Chulalongkorn was well known and widely available, having a portrait made by a craftsman became a popular way to own a copy that was also a unique object. Before starting my discussion of these crafted images, however, I will provide further context on the veneration for the present king and his grandfather.

Promotion of the Royal Image

At first glance, the ubiquity of King Bhumibol may be understood as an expression of the authentic love of the Thai people for their king. Indeed, most of the images also go with the slogan "we love the king" (*raw rak nailuang*), often also rendered as "we [red heart] the king" (*raw [red heart] nailuang*). Thai nationalism has integrated the Hindu-Buddhist view of the king as the apex of society. From this perspective, veneration for the royal image appears as a self-evident continuation of the reverence the Thai have always had for their kings, and the monarchy in general. Yet, the overwhelming presence of the king's portrait is a relatively new phenomenon that accelerated from the 1980s, reaching its zenith only after 2005. This prompts the question of why this love has grown stronger—or, alternatively, why this love increasingly needs to be expressed in public.

The ideology of the Thais' unconditional love for their king is controlled and enforced through a strict *lèse-majesté* law that precludes even the most trivial criticism of the king, or of the monarchy in general (Somchai and Streckfuss 2008). Portraits of King Bhumibol are mobilized to both legitimize and obscure the power of the state and its manifestations. The image of the king functions as a screen to hide longstanding and intensifying societal imbalances and anxieties from view. In the political arena, opinions on core political issues such as redistribution of wealth, corruption, human rights, and democracy are sucked into a vortex of the persistent political antagonisms, but seem to invariably resurface as elaborations of the one slogan: "we love the king."

This is not to say that Thai people are passive consumers of the rhetoric of politicians, the elite, and other powerful stakeholders. The promotion of the royal image has evoked unofficial reworkings, reinterpretations, and unforeseen outcomes. One important example of such a reworking is the sudden popularity of King Chulalongkorn many decades after his death. Like other historical Thai kings, King Chulalongkorn has never been forgotten and has always been revered and commemorated. Yet, in the second part of the 1980s, the reverence shifted into a nationwide personality cult (see Nithi 1993). At the time, Thailand was experiencing major economic and social changes (cf. Pasuk and Baker 1995, 1998). The hopes and anxieties that resulted from the explosion of the Thai domestic consumer market after 1985 made many people turn to supernatural support from holy monks, deities, and deceased kings, among whom the spirit of King Chulalongkorn ranked highly (Jackson 1999a; Nithi 1993, 1994; Pattana 2005; Stengs 2009).[2] The King Chulalongkorn cult may therefore be regarded as one of the many "prosperity religions" (cf. Roberts 1995) that flooded the Thai belief system at the time, "religious movements that emphasize wealth acquisition as much as salvation" (Jackson 1999b: 246; see also Pattana 2008).

That one of the greatest kings of Thai history could become an important source of spiritual support in such mundane and personal issues as taking exams, success in commercial ventures, finding a spouse, or even winning the lottery seems in conflict with official state-ideological representations of Thailand as a modern Buddhist kingdom. Ironically, in Thai historiography it is King Chulalongkorn who is regarded as the founder of the *modern* nation-state. The school history curriculum teaches that he modernized Thai society by introducing modern (that is to say nineteenth-century) technological innovations such as railroads, waterworks, and electricity, and by reforming the administrative and juridical system. Moreover, he is known as the king who saved Siam from becoming a colony—"the only country in Southeast Asia" to remain independent. It is also taught that King Chulalongkorn cared passionately for his subjects: he abolished slavery and traveled into the countryside in person (and incognito) to learn about the ordinary people's real needs. Chulalongkorn is therefore generally remembered as the Great Beloved King (*phra piya maharat*). In this narrative, he appears as the epitome of Thainess. The essence of his perceived achievements lies in the integration of both independence and modernity into Thai identity. The cult of the Great Beloved King is thus not so much about King Chulalongkorn as a historical figure, but about the image of the king as it has evolved in the social imaginary of the Thai people (see Stengs 2009).

The King Chulalongkorn cult reached its zenith in the second half of the 1990s. At the time, portraits of him—the cult's most important material carriers—could be found all over the country, in particular in urban areas. Since 2000, it is clear that while this king is still widely worshipped, his predominant presence has given way to expressions of veneration for King Bhumibol. Images of King Chulalongkorn in public and semi-public realms are gradually fading away. Yet, the history of the presence of the two royal images is closely connected. The relation between the two is bound up with three important royal jubilees in the 1980s: the bicentennial of the establishment of the Chakri dynasty and of Bangkok as the Thai capital, in 1982; the celebration of King Bhumibol's sixtieth birthday, in 1987; and in 1988, the celebration of The Record of the Longest Reign. In that year King Bhumibol broke the record set by King Chulalongkorn, who had been on the throne for forty-two years and twenty-three days.[3] This latter event, in particular, was the impetus for a wide array of texts and images in which the royal qualities of the two kings were compared.

One of the first books to focus on a comparison between the two kings, *Two Great Development Kings* (*song maharat nak phatthana*), was published by the Office of the National Cultural Commission in 1988. The cover picture is a double portrait, a composition that also appears on one of the triumphal arches erected on Ratchadamnoenklang Avenue. Through the stream of television broadcasts, magazines, and newspapers covering the celebrations, the picture must have reached almost every household in the kingdom.[4] King Bhumibol, depicted in color, is placed in the front. He is dressed in a white uniform with golden decorations. In the back, slightly higher and thus evoking the effect of watching over the present king's shoulder, we see King

Chulalongkorn, depicted in sepia tones. The composition suggests that the spirit of King Chulalongkorn is literally behind his—living—grandson, approving and supporting him. By extension, the implication is that although King Chulalongkorn died a long time ago, his spirit is still watching over his country and subjects. The composition has inspired a wide range of variations, to be found in coins, medallions, greeting cards, stickers, and posters. Other portraits may be used, and sometimes prayers or royal symbols are added, but generally the color/sepia distinction is maintained (Figure 2).

It is impossible to distinguish popular imagination from officially promoted images, if only because the official images are influenced by the imagination of the public, and vice versa. Although the celebrations are not solely responsible for the character and scale of present-day devotion to the kings, it seems safe to state that the social

Fig 2 New Year's greetings card depicting King Chulalongkorn (in backdrop) "watching over" his grandson King Bhumibol (in foreground). Photograph: Irene Stengs.

imaginary has been influenced by the jubilees. The celebrations further strengthened the glory of the Thai monarchy as a whole and amplified already existing feelings of veneration for King Bhumibol and King Chulalongkorn. The ample attention for King Chulalongkorn in "The Record of the Longest Reign" celebrations in 1988 may even have given the veneration for this king a decisive impetus for its development into a cult.

King Chulalongkorn Commodities

At the time of my field research (1996–8), King Chulalongkorn's image was omnipresent. A bewildering variety of paintings, photographs, and statues, the portraits are descended directly from the king's own

Fig 3 King Chulalongkorn gift items and mementos. Photograph: Irene Stengs.

awareness of the power of his own image. He was, after all, a *modern* king. He invited European painters and sculptors to the Siamese court, to have himself (and other members of the royal family) portrayed extensively. In the last two decades of his reign, the king's image became "public," something unprecedented in Thai history. His portrait reached the general populace in the form of coins, stamps, picture postcards, and New Year's greeting cards, which were reproduced on a large scale for distribution, at least in urban and elite environments.

The paintings, photographs, coins, stamps, statuettes, and statues of King Chulalongkorn still live on today, as reproductions in the form of posters, visual compilations, stickers, photo books, New Year's greeting cards, statuettes, and larger statues, or as decorations on pieces of china, clocks, wedding mementos, key rings, or other everyday objects (Figure 3). These objects can be bought from portrait shops (selling mainly portraits of kings, monks, and deities), amulet markets, or department stores, at temples or spirit medium sessions, or from door-to-door vendors. King Chulalongkorn objects—that is, objects carrying his portraits—are commodities that cross a wide range of price points, the cheapest objects costing no more than five baht.[5] Apparently, the sacredness of the royal image is not incompatible with mass reproduction.

The explosion of easily available, cheap King Chulalongkorn portraits has inspired a search for the unique. This holds especially for members of the more affluent middle class, who want to create a distinction between their King Chulalongkorn objects and those of others and, by implication, between themselves and other worshippers. One way to create such a distinction is to own an authentic object (that is to say, one contemporary with the king's own lifetime), such as an old coin or newspaper with his image. These are both rare and costly. Yet reproductions can also acquire a degree of authenticity through the addition of "unique value." As noted above, Kopytoff has called this process "singularization," and emphasized that it renders a commodity unexchangeable (1986: 68–75). This is not to say that singularized commodities cannot enter a process of re-commodification. The significance of Kopytoff's argument is the insight that "the commodity is not one kind of thing rather than another, but one phase in the life of some things" (Appadurai 1986: 17). Not only are King Chulalongkorn objects commodities that may become unique; as I will demonstrate, singularization may even be regarded as one of the thriving forces of the cult.

There are various ways to make the sacred special, which often coincide. One of these is the use of specific Northern Thai crafts. By having royal images made by hand, modern middle-class people perform their devotion to the king and at the same time distinguish themselves from the "mainstream." The central case study in my analysis will be a temple in the city of Chiang Mai (Northern Thailand), which I will call Wat Doi Chang.[6] At the time of research, the temple was known (mainly locally) for its life-size gilded King Chulalongkorn statue. Although this object was exceptional, the temple's way of distinguishing itself by creating something extraordinary in terms of size, material, and craftsmanship was part of a wider societal trend.[7]

Showing Making?

Popular sentiments surrounding Chulalongkorn became apparent in the late 1980s. In that period, an increasing number of people came to worship the equestrian statue of the king in Bangkok. In September 1992 a famous Thai movie star, Bin Banluerit, sparked further interest in the cult by declaring publicly (both in the *Thai Rath*, Thailand's most popular newspaper, and on television) that he had survived a terrible car accident thanks to the protective power of an original King Chulalongkorn coin (*rian*), which he wore as an amulet. After Bin's declaration, the number of people paying tribute to the king at the statue increased dramatically, and other centers of worship arose elsewhere in the country.[8]

Although it is unclear whether Bin's media performance might have helped to inspire the development of Wat Doi Chang into a center of the King Chulalongkorn cult, it was in 1992 that the abbot of the temple had a vision of the king in golden attire, with a golden crown, seated on a golden throne.[9] This vision told the abbot to create a gilded statue in its image, and to start a charity project for orphaned hill tribe boys. Thereupon, the abbot began to search books for portraits of King Chulalongkorn, in order to be able to "see what he had seen." It turned out that the scene had been identical to that in a photograph taken during the king's "Second Coronation" on November 16, 1873 (Figure 4).[10] The photograph provided the guidance necessary for the making of the statue.

In my opinion, we have to understand the above creation narrative as a part of the showing-of-the-making of the statue, with the search for the portrait as the most significant element in the rendering of the making process. Had the abbot not recorded his story, one would have assumed that the coronation portrait had been the inspiration for the statue. Yet, the abbot claimed that he was not inspired by this image, but by the vision; in fact, the creation of the statue was not the abbot's initiative, but that of the king himself. The abbot saw the portrait only later, in his search for the right image to fulfill his obligation to make the statue. This "reverse" order of creation adds to the singularity of the statue, giving it an "added" spiritual value.

The statue in Wat Doi Chang indeed resembles the "Second Coronation" picture exactly (Figure 5). Yet, this achievement was not enough to satisfy the abbot, who wanted to guarantee the object's uniqueness. Contrary to what one would certainly think at first sight, the "depth" under the radiant leaf gold "surface" is not a cast image. Its real uniqueness is only revealed at the rear side of the statue. There—the abbot insisted on showing me in person—the craftsman has left a part unfinished, so that we find the proof of its unique quality: it was carved from a solid trunk of teak. Whether it is really made out of a *single* piece of wood must be taken on trust; this showing, or disclosure, is perhaps also a practice of mystification.

Fig 4 Reproduction of the Second Coronation photograph. Photograph: Irene Stengs.

Fig 5 The gilded statue made after the Second Coronation portrait. Photograph: Irene Stengs.

Northern Thai Craftsmanship

The bare spot that shows the making of the temple's statue testifies to the high level of craftsmanship in Northern Thai woodcarving. An important dimension in the self-presentation of Northern Thailand (or Lanna) is that the region's traditional craftsmanship is still alive. In this image of arts and crafts, Northern Thai or Lanna identity appears as distinct from that of other Thais. Paradoxically, it represents at the same time an indispensable part of Thai national identity, providing a romantic view of the nation's past in the form of cultural artifacts. Tourism being an important major source of income, these authentic crafts have developed into a major industry, which appeals to both foreign and Thai tourists (Cohen 2001; Wherry 2006).

Apart from woodcarving, Lanna Thai crafts include paper umbrella making, fan painting, mulberry papermaking, silk weaving, and chiseled metalwork, as well as more ephemeral arts such as dancing and cooking. At annual trade fairs and in villages-turned-handicraft-markets souvenirs are offered for sale, while at the same time tourists can view the artisans at work. Among the teak elephants and furniture, silverware and jewelry, fans, umbrellas, and traditional sweets, there is a prominent place for portraits of historical kings, monks, or members of the present royal family carved in wood. The abbot's choice for a King Chulalongkorn statue carved in teak must be placed in the context of this present-day popular culture, which highly values heritage (*moradok*) and authenticity (*boran*).[11] In this respect, the abbot's choice follows a globally shared, middle-class preoccupation. The making of objects in shops or at markets visually authenticates each portrait. My argument is that this crafting-in-the-open is a ritualized practice that singularizes the end product through the added value of craftsmanship. Yet, this performance of craftsmanship is at the same time a mystification: it deflects attention from the makers and provenance of the other objects for sale in the same place, as well as the earlier stages of the work.

One of Chiang Mai's most popular tourist destinations for foreign and Thai tourists alike is the Night Bazaar. Many stalls sell original Northern Thai crafts. At the same time one can see craftsmen at work in their workshops. For a unique, large-scale, handcrafted King Chulalongkorn portrait, people are willing to spend considerable amounts. In the words of Tui, an artisan at the Night Bazaar, who at the time specialized in portraits of kings chiseled in aluminum:[12]

> Everybody will immediately understand that it is very difficult to make such a portrait. One will also see immediately that one is looking at a portrait of a rare kind. This is because only a few craftsmen are sufficiently gifted to make these portraits. That is why people like to own such a portrait, even if they have to pay a lot of money for it.

Although Tui makes most of his portraits at home during the daytime, his stall at the Night Bazaar also functions as a small workshop. While Tui continues chiseling, passers-by, especially Thai tourists, enjoy seeing him working. The shop displays a few smaller pieces that demonstrate the various stages of the process. Tui's performance of authentic craftsmanship attracts potential buyers, who ask questions about the origin of certain portraits, about the price and the time it would take to make such a portrait on commission. Thus, although Tui told me that the object itself shows its making, he acknowledged at the same time the significance of showing this making. Such "making performances" make the tourist's anticipations come true, so to speak.

Singularization of the Sacred

Both the abbot and the artisan Tui were aware of the importance of the visibility of the making process in the end product. With

Fig 6 The abbot in front of the unfinished Second Coronation statue and a portrait of King Chulalongkorn. The elaborate decoration of roses shows the picture was taken at a Chulalongkorn celebration. Photograph: Irene Stengs.

regard to the performative aspect of the making, however, their positions and interests differ. To clarify this point, I will return to the King Chulalongkorn statue at the temple. When I started my research, the statue had already been made. Yet the abbot had not only ordered the woodcarver to leave a part unfinished, but in order to further preserve the making process he also had photographs made of the statue before its gilding. Most of these pictures are compilations of images of the unfinished statue; others are portraits of King Chulalongkorn or of the abbot himself, sometimes depicted together in a single image (Figure 6). In addition to showing the making of the statue, these compilations point to the spiritual link between the king and the abbot. This is an important dimension in understanding the development of the temple into a King Chulalongkorn cult center. The temple became famous because of the statue's creation narrative, perhaps even more than the statue itself. This story of making offers proof of the abbot's special spiritual connection with the king. This compelling combination not only attracted worshippers, but also generated new instances of singularization through craftsmanship and sacralization.[13]

To place the remainder of this story in a proper perspective, it is important to know that for lay Buddhists the sponsoring of temples in cash or in kind is an important means of earning merit.[14] The donation of auspicious objects, such as portraits of King Chulalongkorn portraits, is one such meritorious act. During the period of research the temple received so many portraits that eventually a separate storage space for them had to be constructed. In general, lay donations of religious images and sacred objects to temples happen on Buddhist holidays, since these days are considered to be the most auspicious. Often such images will undergo sacralization ceremonies, which are also occasions for people to have their own sacred objects consecrated. The ceremonies are also important for the temples themselves. These are well-advertised occasions, both in advance and afterwards. Posters composed of photographs made during one such event—showing the monks and the most important benefactors during the ritual "making process"—testify visually to the import and power of the temple as a producer of sacred objects (Figure 7).

Wat Doi Chang had become an important center for sacralizing King Chulalongkorn objects, and Chulalongkorn Day, October 23, was the ultimate auspicious occasion for such a ceremony.[15] Chulalongkorn Day, the anniversary of the king's death, is an important national holiday. Throughout the country, King Chulalongkorn is remembered in early-morning wreath-laying ceremonies at local administrative centers. At Wat Doi Chang, however, Chulalongkorn Day is celebrated as a religious holiday, and the preferred day to present King Chulalongkorn portraits to the temple, or to bring in one's own objects for sacralization. Any King Chulalongkorn image—whether mass-produced, an authentic original, or a craft product—could enter this sacralization ceremony (Figure 8).

The context of donating as a merit-earning practice explains the lay sponsoring of the carvings of two further over-life-sized statues of the warrior kings King Naresuan (r. 1590–1605) and King Thaksin (r. 1767–1781). Both kings are renowned for successfully defending the Siamese kingdom against the Burmese, the Thais' archenemy (cf. Thongchai 1994). With the three royal statues placed next to each other in the specially constructed Chulalongkorn *wihan*,[16] the temple reflects the strong nationalist sentiments upon which the King

Fig 7 Detail of a poster testifying to the magnificence of an earlier sacralization ritual, Wat Doi Suthep, 2007. Photograph: Irene Stengs.

Fig 8 Objects brought in for sacralization on Chulalongkorn Day 1998, Wat Doi Chang. Photograph: Irene Stengs.

Fig 9 Woodcarver posing with a nearly finished wooden portrait of King Chulalongkorn, Wat Doi Chang, 1997. Photograph: Irene Stengs.

Chulalongkorn cult is built. The statues depict the Northern region as an integral part of the Thai nation. Yet, as authentic pieces of Northern Thai craftsmanship, they also illustrate the region's distinct identity.

Other, more well-to-do, worshippers were inspired to present the temple with a wood-carved King Chulalongkorn portrait or statue. Not long after I had found my way to the temple, a small, temporary workshop was erected, where woodcarvers were working on King Chulalongkorn portraits of over one meter tall, each carved out of a single block of wood (Figure 9). All portraits were copies of well-known photographs or paintings of the king. Eventually, these carved portraits were placed on the ledge next to the abbot's original statue (Figure 10), their size and craftsmanship making them stand out above the average. On most portraits, the donor's name is clearly carved under the image. By patronizing the making of these exceptional (and expensive) portraits, donors distinguished themselves clearly from the mainstream.

Fig 10 Wood-carved King Chulalongkorn portraits with Chulalongkorn Day decorations, Wat Doi Chang, 1998. Photograph: Irene Stengs.

Although the presence of the nearby workplace accentuated the craftsmanship of the portraits, the woodcarvers never demonstrated the early stages of execution; they only seemed to be more or less finishing the portraits. The rough work had been done elsewhere already, possibly by other hands and using modern power tools (cf. Cohen 1998).[17] The workplace, therefore, was revealing and hiding the making process at the same time. Similarly, the pictorial record of the purposefully unfinished King Chulalongkorn statue included no photographs of the actual carving process. The showing of making, as performed in the temple and adjoining workshop, is a mystification and revelation at once—not unlike the performance in Tui's shop and the bare spot on the statue.

What do these various omissions accomplish? For Tui, the performance of his craftsmanship was an important way to authenticate his portraits. It was something he would have to do as long as he was to practice his trade. The temple workshop served a comparable purpose: it stressed the authenticity and craftsmanship of the portraits, and strengthened the image of the temple as genuinely Northern Thai. The unfinished spot on the statue is proof of its unique material properties. The photographic compilations, however, tell a more complex story in which the concrete making is of less importance.

For in addition to visualizing the spiritual connection between king, abbot, and statue, they testify to the making of the statue as a sacred object rather than as a material object. Where sacred objects are concerned, the end of the making process does not coincide with the craftsman's final stroke. It is the abbot's "spiritual craftsmanship" that gives such statues and portraits their finishing touch. And it is this making that the pictures show.

Acknowledgments

The material for this article was collected during my PhD field research on the cult of King Chulalongkorn the Great, which was conducted between September 1996 and December 1997, and in October/November 1998 (see Stengs 2009). The research was funded by the Program on Globalization and the Construction of Communal Identities of the Netherlands Foundation for the Advancement of Tropical Research (WOTRO).

Notes

1 Siam was formally renamed Thailand in 1939. The name of the country was changed back to Siam in 1945, and renamed Thailand again in 1948.

2 The worship of saints—extraordinary venerated monks, living as well as historical—is an important feature of Thai popular religiosity.

3 The Chakri dynasty was founded in 1782, when General Chakri seized power and subsequently made Bangkok the new capital of Siam. King Bhumibol Adulyadej is the ninth Chakri king (Rama IX); King Chulalongkorn was the fifth king of the Chakri dynasty (Rama V).

4 By the end of the 1980s television reached more than seventy percent of rural and more than ninety percent of urban households. The ratings

of the national news are among the highest (Pasuk & Baker 1995: 315).

5 At the time, 1,000 baht was approximately equal to US$25.

6 For reasons of privacy the name of the temple is fictitious.

7 In Northern Thailand, woodcarving is a heritage craft. The hallmark of a temple in a neighboring district, for example, was its large number of giant wooden couches, all with seats cut out of single pieces of timber. In addition, this temple boasted handcrafted statues of the holy monk Somdet To, carved in jackfruit.

8 At the time of my research, thousands of people were coming to the statue every week, particularly on Tuesday evenings, to pay their respects to the king. Tuesday is an important day in the cult. King Chulalongkorn was born on a Tuesday and many believe that every Tuesday night, at 10 p.m., the spirit of the king descends from heaven to enter the statue.

9 The phrase "King Chulalongkorn cult center" is my own. The temple appeared to be one of the most important sources of King Chulalongkorn portraits in the city and vicinity. My rendering of the vision and its results are based on the story as told to me by the abbot. I have referred to this empirical material in other publications (see Stengs 2005, 2009, 2012 [forthcoming]). The following interpretation, however, has been developed for the present publication.

10 The first coronation took place when the king was fifteen years old (1868), and a regent was appointed. King Chulalongkorn was the second king of the Chakri dynasty to have a second coronation ceremony. King Rama I (the former General Chakri) had a second coronation, organized in 1785 when he moved from Thonburi to Bangkok, making the latter the new capital of Siam.

11 The popularity of heritage and authentic Thai crafts and arts is reflected in the increasing number of lifestyle magazines, advertisements, and television drama series situated in a

nostalgic past, and events organized around these themes.

12 The working of aluminum is derived from the traditional chiseled silverware for which Chiang Mai is famous.

13 Emile Durkheim argued that it was only through setting "a certain portion of their environment" apart that societies could mark out sacred space; Kopytoff sees singularization as "one means to this end" (1986: 73).

14 In Theravada Buddhism, merit, karma, and reincarnation are central concepts. A person's karma (*kam*) determines one's rebirth. According to the law of karma, every action generates a consequence. Actions in accordance with Buddhist morals will produce merit and contribute to good karma. The more merit is accumulated the better one's karma, and, accordingly, the better one's present and future existence. Most Thais, therefore, are continuously engaged in a "quest for merit" (Keyes 1983: 267).

15 This is a different sacralization ceremony from the initiation of Buddha statues, which require a so-called eye-opening ceremony (see Swearer 2004). In general, royal imagery does not need explicit sacralization by a ritual expert, but may be charged with further protective powers in Buddhist ceremonies, depending on the owner's needs.

16 The temple building that usually houses the main Buddha statue(s).

17 This farming out of early, not so profitable, stages of the production process is a common practice in craft production (Cohen 1998: 160).

References

Apinan Poshyananda. 1992 (two volumes). *Western-style Painting and Sculpture in the Thai Royal Court.* Bangkok: Bureau of the Royal Household.

Appadurai, Arjun. 1986. "Introduction: Commodities and the Politics of Value." In Arjun Appadurai (ed.), *The Social Life of Things: Commodities in Cultural Perspective*, pp. 3–63. Cambridge: Cambridge University Press.

Cohen, Eric. 1998. "From Buddha Images to Mickey Mouse Figures: The Transformation of Ban Thawai Carvings." In M.C. Howard, W. Wattanapun, and A. Gordon (eds), *Traditional T'ai Arts in Contemporary Perspective*, pp. 149–74. Bangkok: White Lotus Press.

Cohen, Eric. 2001. "Thailand in 'Touristic Transition.'" In Peggy Teo, T.C. Chang, and K.C. Ho (eds), *Interconnected Worlds: Tourism in Southeast-Asia*, pp. 155–75. Oxford: Pergamon.

Jackson, Peter A. 1999a. "The Enchanting Spirit of Thai Capitalism: The Cult of Luang Phor Khoon and the Post-modernization of Thai Buddhism." *South East Asia Research*, 7(1): 5–60.

Jackson, Peter A. 1999b. "Royal Spirits, Chinese Gods, and Magic Monks: Thailand's Boom-time Religions of Prosperity." *South East Asia Research*, 7(3): 245–320.

Keyes, Charles F. 1983. "Merit-transference in the Kammic Theory of Popular Theravada Buddhism." In Charles F. Keyes and E. Valentine Daniel (eds), *Karma: An Anthropological Inquiry*, pp. 261–86. Berkeley and Los Angeles, CA: University of California Press.

Kopytoff, Igor. 1986. "The Cultural Biography of Things: Commoditization as Process." In Arjun Appadurai (ed.), *The Social Life of Things: Commodities in Cultural Perspective*, pp. 64–91. Cambridge: Cambridge University Press.

Nithi Aeusrivongse. 1993. "Lathi phithi sadet pho ro ha" [The Cult of King Rama the Fifth]. *Sinlapa Watthanatham chabap phiset*—special edition of *Art and Culture Magazine* (August), Matichon.

Nithi Aeusrivongse. 1994. "Lathi phithi chao mae kuan im" [The Cult of Chao Mae Kuan Im]. *Sinlapa Watthanatham—Art and Culture Magazine*, Matichon, 15(10): 79–106.

Pasuk Phongpaichit and Chris Baker. 1995. *Thailand: Economy and Politics.* Kuala Lumpur: Oxford University Press.

Pasuk Phongpaichit and Chris Baker. 1998. *Thailand's Boom and Bust.* Chiang Mai: Silkworm Books.

Pattana Kitiarsa. 2005. "Beyond Syncretism: Hybridization of Popular Religion in Contemporary Thailand." *Journal of Southeast Asian Studies*, 36(3): 461–87.

Pattana Kitiarsa. 2008. "Buddha Phanit: Thailand's Prosperity Religion and Its Commodifying Tactics." In Pattana Kitiarsa (ed.), *Religious Commodifications in Asia: Marketing Gods*, pp. 120–43. London and New York: Routledge.

Peleggi, Maurizio. 2002. *Lords of Things: The Fashioning of the Siamese Monarchy's Modern Image*. Honolulu: University of Hawaii Press.

Roberts, Richard H. 1995. "Introduction: Religion and Capitalism—A New Convergence?" In Richard H. Roberts (ed.), *Religion and the Transformations of Capitalism*, pp. 1–18. Oxford: Routledge.

Somchai Preechasilpakul and David Streckfuss. 2008. "Ramification and Re-sacralisation of the Lese Majesty Law in Thailand." Paper presented at the 10th International Conference on Thai Studies, Bangkok, January 9–11.

Stengs, Irene. 2005. "The Commodification of King Chulalongkorn: His Portraits, Their Cultural Biographies and the Enduring Aura of a Great King of Siam." In Wim van Binsbergen and Peter Geschiere (eds), *Commodification: Things, Agency and Identities: The Social Life of Things Revisited*, pp. 299–318. Berlin/Munster: LIT.

Stengs, Irene. 2009. *Worshipping the Great Modernizer: King Chulalongkorn, Patron Saint of the Thai Middle Class*. Singapore: NUS Press in association with University of Washington Press.

Stengs, Irene. 2012 (forthcoming). Portraits That Matter: King Chulalongkorn Objects and the Sacred World of Thainess. In Dick Houtman and Birgit Meyer (eds), *Material Religion and the Topography of Divine Spaces*. New York: Fordham University Press.

Swearer, Donald K. 2004. *The Ritual of Image Consecration in Thailand*. Princeton, NJ and New York: Princeton University Press.

Thongchai Winichakul. 1994. *Siam Mapped: A History of the Geo-body of a Nation*. Chiang Mai: Silkworm Books.

Wherry, Frederick F. 2006. "The Social Sources of Authenticity in Global Handicraft Markets: Evidence from Northern Thailand." *Journal of Consumer Culture*, 6(5): 5–31.

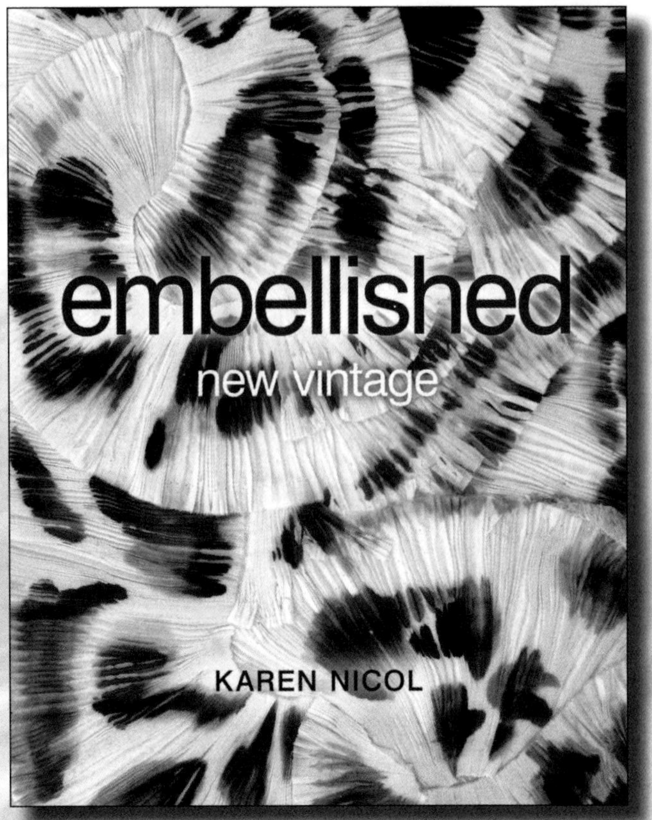

Embellished
New Vintage
Karen Nicol

July 2012 / £30 hb / 9781408105757

Embellished contrasts vintage objects, collected over the years, with cutting-edge contemporary decoration for fashion and textiles. Showcasing a selection of techniques, Karen Nicol explores their potential and pushes materials to their limits to create stunning and unique pieces.

About the Author

Karen Nicol is a well-known British textile artist. Her clients include Whistles, Lulu Guiness and Courtalds. She is a Senior Research Fellow at the Royal College of Art.

www.acblack.com/visualarts

The Journal of
Modern Craft

Volume 5—Issue 1
March 2012
pp. 69–92

DOI:
10.2752/174967812X13287914145514

Material Destinies: Jewelry, Authenticity, and Craft in the American Southwest

Henrietta Lidchi

Henrietta Lidchi is Keeper of the Department of World Cultures at the National Museums Scotland. Her scholarly work is focused on North American collections, and particularly the American Southwest. Her recent publications include articles on representation in museums and Southwestern jewelry, and include *Visual Currencies* (co-edited with Hulleah Tsinhnahjinnie).

Abstract

This article considers the representation and interpretation of Native American jewelry of the American Southwest. It gives a broad history of silversmithing and lapidary traditions, in order to contextualize the tangible and intangible values bestowed on silver and stone as pure, worked, and combined materials. It suggests the benefit of a more nuanced interpretation which teases out the distinctions between indigenous and external attributions of the intangible and tangible values attaching themselves to jewelry as "craft." Considering two iconic photographs taken in the late nineteenth century, it explores differing perceptions of these techniques at this time. While jewelry making has been perceived as part of the ordained quartet of "Indian arts and crafts," its status has been extensively debated under the impact of commercialization and technical modernization in the early twentieth century. Debates that arose in the 1930s, which sought to define and regulate the genuine and handmade aspects of jewelry as craft, raged for much of the decade. This moment of definition has had lasting influence on the perception of jewelry, and the authentication of jewelry as craft, particularly when it is produced for a connoisseurial market. This raises the question of whether viewing jewelry as craft within this context misses a key aspect, namely its function as indigenous adornment, where materials and process are subsidiary to use, color, and communication.

Keywords: jewelry, Native American, silver, turquoise, craft, authenticity, commercialization, trading.

In his introduction to Kathleen Conroy's 1975 publication *What You Should Know about Authentic Indian Jewelry* (Figure 1), the late Lloyd Kiva New, former director of the Institute of American Indian Art in Santa Fe, describes the publication as "a virtual public service document" guiding the consumer, gently and reliably, through the "ticklish" field of Southwest "Indian jewelry" (Conroy

1975: n.p.). New argues that while Native American art is a continuous and indigenous form of expression connected with belief and ceremonialism, the intervention of the market had resulted in imitation and mystification on an industrial scale:

> [T]he amount of Indian jewelry being sold today … [is] … far in excess of what could possibly be produced by hand by the limited number of practicing Native American silversmiths. (Conroy 1975: n.p.)

New voices a persistent concern surrounding "Indian arts and crafts,"[1] namely the debilitating power of an uninformed and seemingly insatiable demand. These comments were written in the 1970s, a boom time for "Indian jewelry" when pop stars were fashionably parading iconic

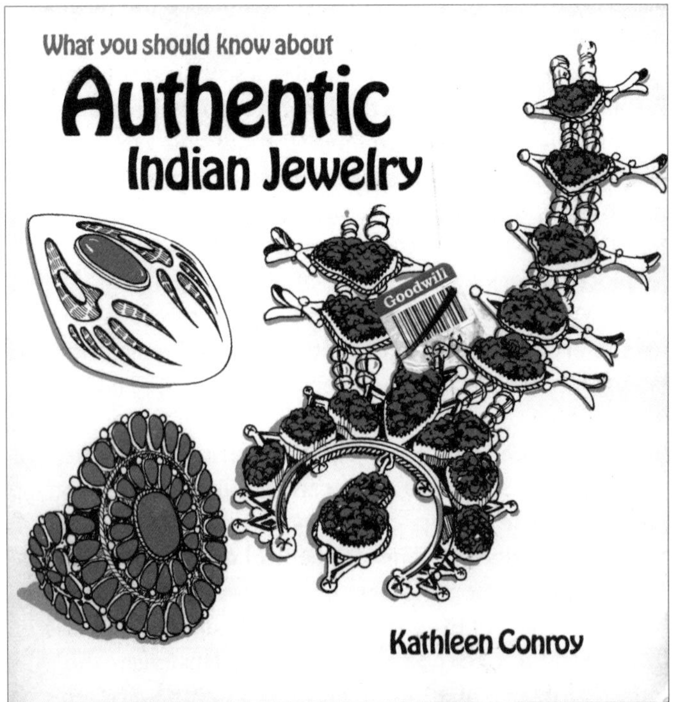

Fig 1 Book cover for Conroy 1975, showing a "high-quality" squash blossom necklace with turquoise nuggets of Navajo style, a clusterwork turquoise and silver bracelet (likely Zuni), and a Hopi overlay buckle or brooch with turquoise cabochon and bear claw design, all iconic styles by the 1970s.

Southwestern styles; but such remarks reflect established debates and could as easily be articulated by writers today.

In the popular imagination, "Indian jewelry" of the American Southwest is understood through the combination of two materials: silver and turquoise (Figure 1). The simple union of metal and stone conceals complex and converging histories: the pre-contact Puebloan lapidary tradition and the nineteenth-century Navajo silversmithing tradition, incorporating the influence of Spanish and American colonists, miners, settlers, traders, and philanthropists.[2] The transformation of the American Southwest from a region that was perceived as hostile in the nineteenth century into the "salutary, educational, heroic, and … ludic region" (Weigle 1989: 115) of the early twentieth set the scene for the emergence of a lively and diversified market for Native American arts and crafts. The propulsion of these arts and crafts into the public arena has repeatedly provoked fears of appropriation, resulting in sporadic attempts at regulation and enforcement in keeping with a desire to visibly preserve, through craft, what might be deemed the "intangible" or inalienable aspects of indigenous culture.

This article considers the representation and interpretation of Native American jewelry as craft. Here it is traced through the use of the two iconic materials silver and turquoise, because these materials (and the processes and forms associated with them) are the means by which Native American jewelry tangibly embodies its links with indigenous cultural traditions. Moreover, it is the manual working of these materials that has been at the core of definitions of authenticity and value by outsiders. Moore

has argued (in this journal) that "the hand … described the very ontology of the Native person" (2008: 209) in the nineteenth and the early twentieth centuries. Moore traces this emphasis through the Silver Hand program developed to authenticate Alaskan Native crafts in 1961. In the American Southwest, as shown by Dilworth (1996), and more recently Batkin (2008) and Bsumek (2008), the definitions and the secondary associations of the "handmade" constituted a major preoccupation in discussions of Native American jewelry during the interwar period. That legacy is carried forward today. In exploring historical debates and their current interpretation, my larger purpose is to situate ideas of authenticity and craft within the broader context of material culture theory. This demands that we see craft and commodities not as oppositional things, but as the material means by which tangible and intangible values circulate within a number of constituencies (Myers 2001).

The Matter of Materials

Native American jewelry is a vital and syncretic craft, but there is straightforward contrast in the histories of lapidary and silver work in terms of longevity. A key attribute of the Southwest is the indisputable fact of continuing occupation over thousands of years. From the third to the fifteenth century, the Greater Southwest (which included current-day Mexico) was home to three flourishing cultural groups: the ancestral Puebloans, the Mogollon, and the Hohokam. These peoples left a rich archaeological record, replete with items of adornment made from organic materials which testify to the artistry, invention, and dedication of makers, each of whom made a distinct contribution to ancient Southwestern jewelry design, particularly in the arenas of beadmaking and mosaic (see Jernigan 1978). Pre-contact turquoise mining, though organized and impressive in its scale, was largely decentralized and according to Weigand and Harbottle "embedded in the dynamics of a trade structure, which operated in service to the overall Mesoamerican population" (1993: 165). Turquoise for much of the pre-contact period was a commodity traded further south, a central component of Aztec mosaics channeled by Mixtec distribution networks (see McEwan, Middleton, Cartwright, and Stacey 2006), although it was clearly part of the local ceremonial complex. The most important turquoise deposits in terms of history and productivity are those of the Cerrillos mining district, twenty-five miles from present-day Santa Fe, New Mexico and near to the pre-contact site of Chaco Canyon.

The conventional history of Southwestern jewelry pinpoints the mid-nineteenth century as the key moment of innovation, when the Navajo first started to work silver. Spanish reports of the 1790s note silver as part of Navajo male apparel, most likely obtained through trade (Wheat 1982: 16) and there are accounts of tin and metal scrap being used for necklaces and bracelets in the early nineteenth century (Kluckhohn, Hill, and Kluckhohn 1971: 300). The use of copper and brass ornaments pre-dates silver, though these materials are more commonly associated with 1930s and 1940s tourist items (see Baxter 2001: 132–4). However, the widespread adoption of silver as adornment followed the acquisition of smithing techniques[3] by the Navajo, most likely as an offshoot of blacksmithing. Silversmithing skills became embedded by the late 1860s, triggered by the rounding up and internment of the Navajo people at Bosque Redondo, New Mexico from 1864 to 1868. Bosque Redondo was a brutal, bitter, and, ultimately unsuccessful military experiment, which resulted in the resettlement of the Navajo by the US Government on a reservation, ushering in significant changes in Navajo culture (Bailey and Bailey 1982: 25–8). The areas where the Navajo clustered close to army and trading outlets, such as Fort Wingate and Fort Defiance, are linked with the earliest practitioners of silversmithing as craft (see Matthews 1968 [1883]). Subsequently, it became a signature craft for Navajo and Pueblo peoples. By the 1880s wealthy Navajos working for Indian Agencies asked to be paid in silver currency, not goods or commodities, which they could melt down and work into jewelry (Bailey and Bailey 1982: 121). The 1880s marks the moment when turquoise beads and earrings started to be set as stones: when materials and traditions combined. Until the 1890s, on the cusp of the commercialization of Southwestern crafts, a Navajo aesthetic of heavy, cast silver jewelry made from coins, worked with chisels and stamps, continued to dominate.

The Value(s) of Materials

Between the 1880s and the early decades of the twentieth century, the production and use of silver jewelry progressed rapidly. Jewelry was opportunistically traded and worn, with silver quickly integrated into indigenous regimes of taste. Jewelry was a necessary aspect of dress at all ceremonial and social occasions, and had a recognized economic value. By the 1880s, the Zuni and Santo Domingo were trading turquoise beads for Navajo silver, and traders based on the Navajo Reservation were taking it in as pawn. For the Navajo, silver was simultaneously a means to dress appropriately and display wealth, as well as a means to accumulate and invest. Jewelry was exceptionally reliable as collateral.

The development of tourism in the Southwest is often cited as *the* economic driver in the transformation of Native American craft practices, from internally orientated items to externally focused consumer products. A much less discussed but clearly influential arena is mining. In the late nineteenth century, mining of minerals and metals prompted unprecedented demographic and economic growth in the newly formed territories and western states. Turquoise and (imported) coral had long been part of intertribal trade and the informal economy of the region, but before

tourism, mining fed the cash economy and supplied raw materials (Figure 2).[4] American and Mexican coins were the primary sources for Native jewelry makers[5] so the development of the craft was intertwined with fuller access of Native people to money, and better integration into the cash economy (see Bsumek 2008: 48). By the 1930s the source and quality of the silver used for Navajo jewelry had become a pivotal consideration. At this juncture histories of the region (both pre-contact and more recent) were marshaled into an argument concerning authenticity as turquoise, and more especially coin silver, became the *de facto* "raw" materials that required the craftsman's hand to transform them. This involved an obvious paradox. Casting money for jewelry unambiguously denotes its *tangible* exchange value, even though, as we shall see, the use of coin silver was the crucial signifier of *intangible* cultural value until the 1940s.

Writers commenting on turquoise in the context of Navajo and Pueblo culture place

Fig 2 Sleeping Beauty mine shop selling turquoise, Globe, Arizona, February, 2007. Photograph: Henrietta Lidchi.

it high within the hierarchy of significant materials that hold ceremonial, economic, and aesthetic value.[6] Witherspoon, for example, asserts that, to the Navajo, turquoise belongs to the domain of *ntł'iz*, which he defines as "rigid materials with ceremonial value," a classification that also includes abalone, white shell, and jet (1981: 47; see also Franciscan Fathers 1968: 411). An integral part of Navajo ceremonials, turquoise actively maintains *hózhó*: the state of harmony, beauty, and well-being toward which all holy, animate, and inanimate beings in the Navajo universe strive (Witherspoon 1977; 1981). While the indigenous symbolic value of turquoise can be considered to be innate, applicable to the raw material, as well as the worked stone, silver comes to hold aesthetic and symbolic value only through manufacture. It acquires the properties of *ntł'iz* only when associated with turquoise; otherwise it is white metal, or *beesch łighai*, appreciated for its hardness and ability to take on sculptural form (Witherspoon 1981: 47–8). So silver has the power to actualize, maintain, or restore *hózhó* only when it has been subject to the process of creation and transformation, into a pleasing and harmonious piece of adornment (Witherspoon 1977: 151).

To acknowledge jewelry as a locally embedded while commercially significant craft is to appreciate that different values are conferred on turquoise and silver as pure, worked, and combined materials. These dual relationships suggest the need for a more nuanced interpretation which teases out the distinctions between indigenous and external attributions of the intangible and tangible worth attaching themselves to jewelry as "craft."

Visualizing Native Crafts

For a formerly ignored area of indigenous creation to be identified and distinguished as "high art" or "craft" requires both beneficial conditions and facilitating biographies. Writing about the market for Australian Aboriginal art in the 1980s, Morphy notes the speed with which it moved from the periphery to the center of national imagination and identity, arguing that this was the result of two "acts of creation" (1995: 213). The first involved bringing the object into being—a physical and conceptual act pertinent to the communities of creation. The second was its transformation from object into commodity, no less a creative and conceptual act, but one orientated away from the communities of creation toward "consumers," both nationally and internationally. The sudden prominence of Australian Aboriginal art required new forms of recognition and channels of consumption. In the American Southwest, Native American arts and crafts underwent a similarly accelerated process of exposure and acceptance, albeit a hundred years earlier.

The intensification of ethnographic and archaeological interest in the Southwest in the late nineteenth century was directly and indirectly influential in nurturing Native crafts and the fledgling market. It generated imagery of the area and its peoples (see Dilworth 1996; Mullin 1996; Wade 1985). In considering the visualization of craft at this early stage, and the archetypes of the craftsman that emerged as a consequence, two portraits come immediately to mind: Slender-Maker-of-Silver (Figure 3) and Governor Ahfitche (Figure 4). These portraits of a silverworker and a lapidary are intimate viewings of the spectacle of

Fig 3 "Slender-Maker-of-Silver" or Beshtlagai-ithline-athlsosigi, taken by A. Frank Randall or George Ben Wittick (c. 1885). Courtesy of National Anthropological Archives, Smithsonian Institution (Inventory Number, 02274600).

craft (see Dilworth 1996: 145–6) and were taken respectively at Fort Wingate or Fort Defiance on the Navajo Reservation[7] and San Felipe Pueblo.[8] The comparison here is between two posed photographs, encoded as Navajo or Pueblo through articles of clothing and adornment, as well as the skills being vouched for or demonstrated. There

are visible backdrops and props (see Johnson 1998), including the ubiquitous blankets and the wrapped headbands, adopted later as signifiers of the Native craftsman (see Bsumek 2008: 180–1).

Of the two, the portrait of Slender-Maker-of-Silver (Figure 3) is the more unusual and distinctive in terms of composition. The subject looks directly at the camera. His posture is open, his attitude direct and assured. Wearing jewelry, Slender-Maker-of-Silver returns the viewer's gaze and offers to it a substantial *concha* belt which symbolizes the pinnacle of his labors, although the bridle is a similarly impressive testament to his superior skills. The quality

Fig 4 Stereocard of "Governor Ahfitche" or José Quivera, governor of San Felipe Pueblo, taken by J.K. Hillers (1880). Courtesy of National Anthropological Archives, Smithsonian Institution (Inventory Number, 00230700).

of his workmanship is further confirmed by the well-placed leather pouch, revealing carefully positioned silver bangles. Governor Ahfitche (Figure 4) on the other hand is a more preoccupied figure, absorbed in the mechanics of doing: he is demonstrating the method of making a bead, presumably of turquoise, with a pump drill. Both images connote the innate romanticism of craft, something deriving from the informally honed skill and traditional relationship to material which, by further suggestion, determines form and technique.

As Dilworth (1996: 192) has argued, photographs such as these create timeless images of "primitive" labor, acquiring first iconic and then popular value. These encoded images are *myth* in pure form: "what the world supplies to myth is a historical reality; … and what myth gives in return is a *natural* image of that reality" (Barthes 1972: 155). In them, silversmith and lapidary incarnate the nobility of work made by hand. Although depicted alone, they craft "the work and themselves, not in alienated isolation but with a clear understanding of their purpose and place in the universe" (Dilworth 1996: 145). As artisans, they are independent, embedded in communal relations and morally authentic. Craft here is supplemental to indigenous culture, springing from it, not in dialog with it (Adamson 2007: 11–13) and indigenous culture is, by connotation, instinctive, essential, and unchanging. The slippage between identity and craft appears seamless (Navajo and silver; Pueblo and turquoise).

Materials and Perceptions

There are, nevertheless, fine distinctions to be made between these portraits that offer something of a comparative view of silverworker and lapidary. The portrait of Slender-Maker-of-Silver presents him resolutely as a craftsman. His work is recorded and identifiable; there is a fixing of authorship to the luminescent belt, conspicuously held as a made item, and the bridle above the maker's right hand. However, the *punctum* (Barthes 1984: 57) is his face: it is this that first and foremost captures attention. Charles Frederick Holder, in his 1906 article for *The Craftsman*, provides apt commentary on this early image:

> These people are skilful makers of jewelry, pounding and hammering bands and bracelets of silver out of silver dollars, and displaying no little taste in their designs. It is not the jewelry however, which attracts one's attention, but rather the strong face of the worker—a type long to be remembered. The Navajos are amongst the finest specimens of all the American Indians to-day. (Holder 1906: 752)

Governor Ahfitche, in contrast, has a more elusive presence, the eye more clearly drawn to his hand and the pump drill. This is a sparer image: there are few tools, no jewelry, just the slightly hunched lapidary. The eye is drawn to the persistent "archaic" technology, which, for this pose, he uses uncharacteristically to create a bead.[9]

These visual characteristics suggest differences in outsiders' perception of these two jewelry techniques. Silversmithing in the 1880s was a novel and innovative Native craft, involving a material whose value was appreciated and on the ascendant (see Venables 1995). Lapidary work on the other hand was being revealed, through archaeological excavation and contemporary ethnography, to be an ancient and stylistically continuous craft which used shell and turquoise, a stone of, as yet, little recognized worth. George Frederick Kunz, the gemologist who popularized American stones,[10] observed the Native use of turquoise while remarking on its wider potential for mainstream jewelry. Kunz noted the symbolic role and importance of turquoise in New Mexico. He wrote of the "desultory manner" in which contemporary local lapidaries and Pueblo people mined at Los Cerrillos (Kunz 1890: 57) and the poor quality of stone sold at low prices as souvenirs.

From today's standpoint, there seems to be a different order of subjectivity and agency being constructed by the camera. Slender-Maker-of-Silver has acquired an identity and pedigree as an ancestral master silversmith; he has subsequently been written into the history of the craft. Bedinger (1973: 18–19), for example, credits him as the best early Navajo silversmith and declared his work made between the 1880s and 1890s as the pinnacle of Navajo classicism, coming after early experimentation and prior to the ills of commercialization. In keeping with this attribution there have been continuous attempts to uncover and collect his work. No such lineage is accorded to lapidary work of the same period. Governor Ahfitche, while demonstrating a craft which remains very highly valued, necessary, and sought after by Pueblo and Navajo people, seems to be a cipher, channeling without mediation what has often been written about as an essentially immutable craft practice.[11] Some of these distinctions may arise because the classic techniques used by Native lapidaries and silversmiths have differing potential

for retrospective personal and communal identification. Silversmiths' work can be traced through the use of individual stamps and styles. Traditional lapidary work, such as bead making and tab necklaces, allows identification only through communal styles and likely provenance.

Produced for ethnological and commercial purposes,[12] these two images circulated widely as visual archetypes— representations of "Indians" and "craft." In the "crowded visual marketplace of nineteenth-century America" (Sandweiss 2002: 54), they were sold as reproductions in the form of cabinet cards, *cartes de visite*, and stereographs. Subsequently the visual language of the lone craftsman (jeweler), head down and absorbed in his work, has gained the greatest currency (see Dilworth 1996: Fig. 20, p. 93) (Figure 5). Such is the dominance of this representation that it is fertile ground for contemporary art and commentary. The artist Marcus Amerman (Choctaw Nation of Oklahoma) has recreated and recirculated the portrait of Governor Ahfitche as *The Bead Maker* (2005), as an image in beads (Figure 6). This seems to signify both the continuance of craft and the symbolic entrapment or "metonymic freezing" (Appadurai 1988: 39) that has become part of the mainstream understanding of Southwest craft.

Valuing Craft

To return to Morphy (1995): in discussing the prominence of Aboriginal art in Australia he argues that the two acts of creation bringing new art to the fore (first as an object, second as a commodity) are separate in time, space, and concept, although market forces seek, ultimately, to align them.

Fig 5 Shops along Historic Route 66 in Gallup, New Mexico, showing characteristic advertising rhetoric and imagery, August 2010. Photograph: Henrietta Lidchi.

Here lies the double bind of the market. Successful alignment is inevitably interpreted as contamination, generating a reactive need for distinction. So, though Morphy notes that there was clear Aboriginal agency in creating and sustaining the market that emerged in the 1980s, he nevertheless states that artists came to occupy contradictory and sometimes powerless positions, particularly as regards definitions of authenticity, which were determined from outside. The valuing of art and its popularity depended on the aura of "primitiveness," so that it was highly regarded only in so far as it was "pristine, primeval and, as such, liberating" (Morphy 1995: 214).

At the beginning of the twentieth century the popularity of "Indian arts and crafts" owed much to the American Arts and Crafts movement and the aesthetic reform it promoted (Dilworth 1996; Herzog 1996; Hutchinson 2009). Lapidary work and silversmithing were categorized as part of the quartet of ordained[13] "Indian arts and crafts," alongside basketry, textiles, and ceramics, but were seemingly less prominent in craft writing or in the minds of collector-connoisseurs. At this early stage, they were not as automatically proffered as examples of artisanal utility and "symbols of a simple life" (Herzog 1996: 82) as Pueblo pottery, Californian and Southwestern basketry, and Navajo weaving. Native American crafts retained their mystique into the Depression era. They continued to be associated with an authentic, indigenous America and were caught up in the search for regional identity and a reverence for a simpler, more spiritual

Fig 6 *The Bead Maker*, Marcus Amerman, 2005, glass seed beads sewn over a photograph. Courtesy of the Heard Museum, Phoenix, Arizona, Cat. No. 4407-4.

past (Mullin 1996: 169; Steiner 1983). By this time, as a consequence of its popularity, portability, and commercial success, jewelry enjoyed a much more prominent, if somewhat ambivalent, status. Fifty years after the first systematic anthropological description of Navajo silversmithing as a craft (Matthews 1968 [1883]), Raymond Ortis, writing for the Southwestern Indian Fair, provides a commentary on the impact of market forces:

> Silver spoons, ink wells, paper cutters, trays and numberless articles of trash bearing semi-Indian designs which

are found on sale nowadays may be Navajo in manufacture, but are certainly not so in feeling or conception. They represent nothing more than the abortive adaptations of Indian craftsmanship to American commerce. (Ortis 1935: n.p.)

Drawing a distinction between authentic Navajo articles (brides, bracelets, bead necklaces, etc.) and those that may be "artistic adaptations," he notes that in his view much contemporary work is neither skillful nor artistic, "nor by any stretch of the imagination, Indian" (Ortis 1935: n.p.). Ortis's cautionary stance, which is only taken in relationship to jewelry manufacture, is nevertheless characteristic of those prominent "well meaning and wealthy people"[14] who sought to influence, regulate, and authenticate the production of "Indian arts and crafts," and protect them from the evils of commercialization. By the 1930s philanthropic humanism and benign government intervention were actively trying to "revive" and regulate Native American crafts through a raft of associations, art fairs, and other initiatives while simultaneously advocating against systematic market incursion (McLerran 2009: 73–84; Mullin 1996; Schrader 1983; Wade 1985). Erna Fergusson, commenting on the frenetic activity and philanthropic zeal of Santa Fe "Society," remarked witheringly: "Everybody has a pet pueblo, a pet Indian, a pet craft" (1936/37: 377).

The Regulation of Craft: Man versus Machine

Perhaps no instance illustrates the fears about commercialism and the need to reify the uniqueness of Native jewelry more clearly than the dispute between the Federal Trade Commission (FTC) and the Maisel's Trading Post Co. (Maisel's), of Albuquerque, New Mexico in the early thirties. The FTC regulated interstate trade and had investigated other companies, but in 1932 it brought a case against Maisel's on the basis of misrepresentation supported by a coalition of philanthropists and Southwestern traders[15] (Batkin 2008: 183–6; Bsumek 2008: Chapter 6). Maisel's had entered the curio trade in 1923 and correctly assessed the general consumer's affinity for lighter, inexpensive jewelry sold as "Indian made" (Batkin 2008: 129–31). Maisel's advertising capitalized on the prevailing perceptions of the Navajo craftsman, giving them a competitive edge. This was due to the association in the minds of consumers between "Indian made" silver and its connotation of work that was "handmade" according to Navajo traditional methods and techniques (Bsumek 2008: 178–87). However, Maisel's "Indian made" jewelry was created mostly by Pueblo, not Navajo smiths, and from "commercially produced silver, both coin and sterling" under Euro-American overseers in partially industrialized conditions (using machinery and bench-based processes) using turquoise that was pre-cut to a set number of sizes (Batkin 2008: 131, 183–6).

FTC hearings were conducted in a number of towns across the Southwest and thousands of pages of testimony were submitted to a Board of Review.[16] Its conclusions were clear: mechanization tainted the craft, changing the economics of production. Maisel's were perceived as disadvantaging the more traditional craftsman, generating unfair competition,

creating a bad example, and damaging consumer confidence and goodwill.[17] Stating that 10 percent of Navajo smiths and 20 percent of Zuni smiths depended on hand labor for survival, the FTC found that the future economic security of Native communities and the curio trade were being endangered.[18] Drawing on expert testimony, the FTC argued the case for a craft whose uniqueness was being compromised:

> The individuality, the artistic quality and the beauty of the article is lessened to the extent that machine work replaces hand work in any process of production of silver jewelry. The fashioning of an article of jewelry by the hand application of a hammer is a process of creation into which the silversmith puts his individual powers as a craftsman and artist. The substitution of mechanical rollers … diminishes greatly the quality of the product.[19]

Testimonies especially identified the pounding of coin silver as *the* distinctive Navajo technique, and *the* signifier for authentic Native self-expression. So much so that witnesses testified that hand-hammered slugs made from Mexican coinage (*pesos*) were identifiable to the expert eye and preferable in terms of "temper and durability."[20] Mechanical rolling tainted the result because, as asserted by the anthropologist (and popular author) Oliver La Farge, modifying coinage through pounding invested the metal with "soul" (quoted in Bsumek 2008: 186). Moreover, this perception created an important romantic attachment in the buyer.[21]

Maisel's provided a spirited defense. It professed to have no view as to the soul of the artist[22] and sought to prove that there was no substantial difference between jewelry hammered from ingots or that made using mechanical rollers. It claimed there was no virtue in submitting Native jewelers to "mere drudgery and wasted effort."[23] Maisel's appositely commented:

> It appears that the laborious beating out of the silver from a Mexican dollar or *peso*, or from the slug, has no relation whatever to the finished product unless it be that those few select and artistic souls who fancy the idea, are more pleased with the thought of the difficulties of the craftsman than with the results of his labors.[24]

Maisel's distinguished two kinds of silversmiths: those on the Reservation and those trained at Indian schools, who lived more urban and mobile lives.[25] Modern techniques and working for wages did not in and of themselves mean that the urban-based Native jewelers had "forfeited their birthright to be called Indians," but rather that they earned the right to "have the work of their hands marketed and sold for what it is."[26]

Although Maisel's was accused of siphoning off custom from Reservation-based traders and traditional silversmiths, the niche it was manifestly targeting was the curio market (Bsumek 2008: 183), especially the lightweight jewelry produced by H.H. Tammen Co., whose rhetoric of "Indian design" signaled its distinction from "Indian handmade." Tammen was one of the most successful curio dealers in the West (Batkin 2008: 15) and desirous of distancing itself from competitors. In his statement to the FTC, Carl Litzenberger, president of Tammen, characterized the company's jewelry as

"practical jewelry" that satisfied souvenir needs, in contrast to heavy, traditional jewelry aimed at the Navajo consumer or craft connoisseur. He defensively referred to the changes in jewelry production on the Reservation, noting that Native jewelers had discontinued the use of coins and that silver was "almost entirely supplied today through traders and banks with rolled out coin silver bar or plate," while turquoise was acquired pre-cut from manufacturers or wholesalers.[27] Litzenberger argued for the legitimacy of Tammen's labeling ("our Indian Design jewelry is produced by American workmen"[28]), and as a line of defense, invoked the economic background of the Depression. In private correspondence he later commented, "all of these things are brought about from the fact that nobody is doing enough business to make both ends meet."[29]

The Regulation of Craft: Market versus Maker

Against the background of this very public discussion of authenticity, manufacture, and circulation, papers held by the Fred Harvey Company spanning the 1930s reveal the views of those involved in the jewelry trade and provide a more private and candid commentary as to prevailing business practices and popular taste. The Fred Harvey Company was instrumental in developing the Southwest as tourist destination, and by 1900 was one of the main jewelry manufacturers in the region, with workshops in Albuquerque employing Navajo smiths and further outsourcing of materials to Navajo smiths on the Reservation (Batkin 2008: 112–15; Falkenstein-Doyle 2011: n.p.).[30] The firm's papers, now in the Heard Museum,

Phoenix, include a vigorous correspondence between Herman Schweizer, who managed the Indian Department and who had jobbed out materials to Navajo silversmiths (Moore 2001: 23), and the more cautious J.F. Huckel, who had created the Indian Department and was the Company's vice-president.

Huckel wrote in 1932 that jewelry was the second most lucrative Navajo business,[31] arguing that "phoney, imitation Navajo silver"[32] debased the market, jeopardizing the possibility of Navajos making an honest living, and should not be stocked. He described the Harvey Company, by contrast, as a humanitarian enterprise dealing in "genuine" jewelry whose self-appointed role was to stop curio dealers and traders from stealing the livelihood of "innocent and ignorant Indians."[33] Schweizer issued a lengthy rejoinder, asking at the same time whether Huckel was advocating taking Tammen jewelry off display.[34] Describing the history of the curio trade, he argued that the Harvey Company was necessarily invested in the tourist market: "there is a demand for Indian or semi Indian things as souvenirs from the West which I do not think can be ignored."[35] He identified certain compromises as necessary to satisfy popular tastes, because "genuine" Navajo jewelry could not be made at competitive rates:

> [T]he Navahos cannot successfully make in quantities light weight brooches, earrings and many other small items. Their primitive tools are too crude for things of that kind.
>
> We have attempted, at great pains, for over 25 years to have these brooches, earrings and all that sort of thing made by the Navahos, and even after they come

from the Reservation brooches had to be finished by our Indians here, as they were too crude and then a native Mexican had to put on the brooch back or pins. At best, when finished, these articles cost us 50% more than the copies or imitations could be retailed for.[36]

Schweizer had candid relations with curio dealers and noted that Maisel's model would likely be imitated by others, changing the means of production of Indian jewelry for good. Furthermore, he remarked that improved education and other economic opportunities were making "genuine primitive hand-made work" hard to find, adding that the formal teaching of craft in Indian schools as advocated by Government reformers or philanthropists produced jewelry "no more Indian than imitations made outside. That is, it isn't primitive work."[37] It was "difficult to tell the difference," he wrote, between rolled, sheet, and hammered silver in any case. In short, his message to Huckel was: modernization was coming, no matter what.

The debate around Maisel's and the term "Indian made" simmered on for many years (1937 is the last mention in the FTC Annual report), with Federal and other initiatives emerging under its influence hoping to authenticate and label "genuine" Indian jewelry (and other crafts), in the belief that stricter policies and assiduous promotion would improve the welfare of Native communities (Bsumek 2008; McLerran 2009; Mullin 1996; Schrader 1983). Some dealers felt that this just amounted to "messing around with the curio business."[38] In July 1933, the Secretary of the Interior prohibited the sale of imitation jewelry in National

Parks, causing Schweizer to comment, "the new deal is going to work like a steam roller on a lot of things."[39] The Federal Trade Commission published its findings in August 1933 and issued a cease and desist order, confirmed in 1935.[40] This articulated more specific restrictions surrounding the attribution of "Indian made" (see Batkin 2008: 184–5) and the FTC's desire to regulate rolling, pressing, or ornamenting by machinery. According to the FTC, the use of silver ingots made from coins devalued the craft, whereas pounding pre-made slugs of the "general fineness"[41] of coin silver did not. Buffing machines were accepted in so far as they were not considered "means of production."[42] By 1936 one of the first tasks of the New Deal agency, the Indian Arts and Crafts Board, was that of establishing and publishing "Standards of genuineness and quality for Navajo, Pueblo and Hopi silver and turquoise products" approved by Government stamp[43] to be applied to "truly hand-fashioned and authentic Indian silver and turquoise products."[44] These regulations concentrated on process and materials, including hand-hammering silver, creating silver dies from scratch, and using uncolored, hand-cut, and polished turquoise (where polishing wheels were driven by hand or foot)[45] (see Batkin 2008: 198–9). These issues were inextricably linked to those of discernment, and the self-appointed authenticators of craft applied the new standards with varying degrees of effectiveness. The continuing emphasis on the handmade was an attempt to guarantee the intangibility of jewelry as craft through methods and materials, integrating useful modernizations such as polishing by insisting that this was allowable since

it was undertaken "after the jewelry has been entirely completed."[46] The manifold initiatives of the government agencies, though well meaning, were over-specific, largely unworkable, and in some instances (unsurprisingly) met passive resistance (see Falkenstein-Doyle 2011: n.p.). They foundered by the beginning of the 1940s, at the point when silver itself was no longer accessible to mainstream jewelers.[47] After the Second World War the emphasis on coin ingot and uncut turquoise decisively receded, as some jewelers moved into a studio mode.

Predictably, the rarer voices at this time were those of the Native American jewelers themselves, but when their perspective was recorded they articulated a general concern for quality, conditions, wages, and self-determination[48] while admitting to a willingness to modernize their craft and expressing some cultural indifference as to the importance of hammering out coin silver (Batkin 2008: 198–208; Bsumek 2008: 193, 205–7). Like many other Native American craftsmen in uncertain political and economic times, they participated in the arts and crafts business to ensure themselves a livelihood, and in so doing worked "to invent the modern Southwest and to find a place … within it" (Moore 2001: 23). Indeed, one could posit that by giving tourists what they wanted, these jewelers were attempting to preserve their most treasured handmade possessions for themselves as heirlooms and pawn (which moved in and out of circulation), since these were the very items that connoisseurs so avidly desired.

Craft after Virtue

The publicly polarized discussions of the 1930s expressed the split affinities of the age, with agencies and individuals debating the future of "Indian made" jewelry with passion and conviction. Those promoting cultural revivalism in a period of financial stricture placed crafts within regional traditions and a nativist discourse. These voices disputed the value of craft with their opponents, who embraced mechanization and the progressive forces of commodification as a sign of modernism; yet both had definable vested interests. Eventually the celebration of "genuine" and "primitive" artisanal production methods proved incompatible with market forces, developing technical opportunities and the desires of Native artisans themselves. The ideal of the lone craftsman working outside the reach of the curio trade, diligently manipulating coin silver and hand working turquoise, was clearly at odds with the reality of the modernizing Southwest (Bsumek 2008: 190–3) as well as the transformation in the manufacture of silver goods more generally (Venables 1995: Chapter 4).

Nevertheless, debates from the thirties put into place a relatively enduring framework for describing "authenticity," and therefore the intangible value of Native American jewelry. Those seeking to analyze and describe the far more complex and diverse market that exists today have added a third term, "art," to craft and souvenir, adding a new level atop the hierarchy of jewelry production (Gilster 1999). Indian Market in Santa Fe, the largest judged show and selling event for contemporary Native American art in the USA, publishes yearly standards guidelines. Today more than a thousand artists submit their work for judgment and sale over a long weekend in late August known locally as "Market"

(or cynically as "Markup"). The organizing body, the Southwestern Association for American Indian Arts, has roots in the Southwest Fair of the 1920s and the initiatives of that time which sought to create a direct and authentic link between craftsman and purchasers. The juried show and Market itself continues to showcase indigenous arts and crafts, distinguishing them from those of more affordable, often imitative, work sold outside its boundaries. Participating artists, a significant proportion of whom are jewelers, submit their work for judging and undertake to sell it in accordance with published standards (Figure 7). Each year these are reviewed and renegotiated, with criteria introduced and/or removed to regulate eligibility for submission. Reissuing and ensuring standards in jewelry is acknowledged to be especially taxing, because of the medium's popularity and the inventiveness of those who seek to mimic natural stones and traditional techniques (Bernstein quoted in Russell 2009). In the past, standards for "traditional" work were more restrictive than those for "contemporary" jewelry. However, as jewelry makers are embracing new materials and influences, the implication that craft can be fixed to moments in time deemed more "authentic" and "traditional" has proved increasingly unpopular and untenable (see Jackson 2010). Consequently, in 2010 standards did not produce hard distinctions between traditional and contemporary jewelry. This was consistent with the view that such regulated forms of "authenticity" crush innovation, and that too great an adherence to materials and techniques restricts the field of Native American jewelry as a form of artistic expression.

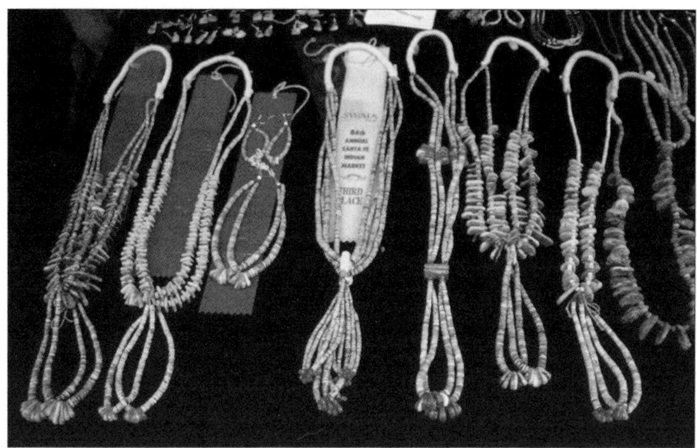

Fig 7 Range of prizewinning bead and tab necklaces in a variety of traditional styles, of natural Southwestern turquoise and spiny oyster shell. Made by Ray Lovato (Santo Domingo), Santa Fe Indian Market, August 2005. Photograph: Henrietta Lidchi.

This is a fluid situation, made even more so by the plethora of imitations of Native made jewelry and Native design currently on the market. However, the continuous desire to formulate indices of authenticity points to two important factors relating to Native American jewelry in the Southwest. First, dynamic markets create excess and contradiction in the meaning of items and the exchanges of which they are part. This may mean that those supposedly shared values are understood differently, depending on what is being consumed and who is doing the consuming. Keane has noted, "market economies do not do away with inalienables [intangibles] so much as re-order the regimes of value in which they function" (2001: 66). Second, the function of jewelry is that it should be worn, and wearing jewelry is a social and personal act. It presupposes, and animates, a visual debate between maker (or giver), wearer, and viewer (Goring 2006). Native American jewelry is visible adornment, and symbolic value is ascribed to its use, style, color, and materials. There may appear to be a concordance between Native beliefs, where process and materials (rather

than product) are aesthetically and spiritually pre-eminent, and the zeal of the 1930s reformers; but this bypasses a significant fact. The intangible qualities of jewelry are not restricted to making or materials; rather they are enacted through display and transactions and registered through color. For example, although the process in which turquoise and silver jewelry is locally implicated—trade, pawn, or sale—involves tangible value, this economic significance does not strip it of potential to mobilize inalienable attributes. As a signifier of cultural identity and tradition, jewelry remains as powerful today as in the past, even though materials, styles, and sources have undoubtedly changed (Figure 8). Neuman highlighted this in the 1950s, noting that while connoisseurs sought out bold and plain silver—"authentic" Navajo jewelry—this antiquarian taste was at odds with contemporary Navajo preferences, for flashier Zuni clusterwork, with its abundant use of pre-cut and reasonably priced turquoise (1971 [1950]: n.p.).

As a transacted object of art and culture, jewelry in the Southwest is a vehicle for self-reflection and projection (Figure 9).

Fig 8 Hopi dancers with turquoise and silver jewelry watched by others wearing jewelry, Saturday morning parade, Gallup Inter-Tribal Ceremonial, August 2005. Photograph: Henrietta Lidchi.

Consequently, notions of authenticity and value need to be broadened to encompass the more playful attitude to materials, styles, and processes prevalent among Native American consumers, particularly as regards the mimicry of turquoise. We should also be alert to the visual dialog that emerges through making, wearing, and viewing. In the Southwest, artist and wearer are not separate identities: making and using jewelry for traditional reasons is part of life. Indeed, sometimes jewelry is so descriptive and figurative that the wearer can literally be wearing a miniaturized and idealized image of him/herself. As the jewelers Gail Bird (Santo Domingo/Laguna) and Yazzie Johnson (Navajo) recently remarked, it is vital to dress right and have appropriate jewelry for communal and ceremonial occasions: "You have to have jewelry. That is why the art continues" (2011: n.p.).

Conclusion

This paper has briefly explored the myth of the craftsman and reviewed histories and debates concerning materials and manufacture. It considered the tensions that emerged in the 1930s revolving around market, manufacture, maker, and collector,

Fig 9 Float at Shiprock Fair Parade adorned with outsize bracelets, belts, and other pieces of jewelry, Navajo Nation, September 1998. Photograph: Henrietta Lidchi. Courtesy Trustees of the British Museum/Henrietta Lidchi.

arguing that these have helped define the course of Southwestern jewelry. Manufacture and materials are still, today, the basis on which judgments are made concerning the uniqueness and authenticity of Native American jewelry as craft. Moreover, the attempts to codify authenticity, which seek to stem and keep up with unfettered mimicry through listing acceptable materials, techniques, and design, is an ongoing phenomenon. The popularity and enduring relevance of jewelry as a craft has necessarily meant an interchange of ideas concerning value. Ultimately, makers today, as in the past, need to choose between economic strategies: making more for less, or less for more, the meaning of which, in a vital and capricious market, is frequently open to question.

It is an oft-repeated caricature that anthropologists are on a relentless quest for meaning and symbolism. Here they share a narrowness of vision that echoes those of craft reformers, who had similarly unrealistic desires for authenticity, eliciting bemused and indifferent responses from those they seek to question. Like all caricatures this one is based in truth, and I have a fond recollection of myself in this guise. In 1997, early in my research, I interviewed a Navajo jeweler in his late fifties in Shiprock, on the Navajo Reservation. He took up the craft relatively late, having been taught by a friend. In his words, jewelry was a viable alternative to a career in blues music. I asked him about the symbolic content of his popular pieces, and he responded:

> It's up to the individual who makes it. If I make it for, like for you, then I put something on there that … connects us, you and me, and I would tell you about

> it … If I am making something for you, then I'll do that, but other than that it's mostly if it sells I'll make more, if it doesn't, I won't make any more. Just economics, just supply and demand. Isn't that what they call it?

This was, obviously, superficially disappointing. Ultimately, though, it helped me realize that the transactions surrounding jewelry happening all over the Southwest do have, more generally, a meaningful part to play. Experiences like this one orientated me away from finding symbolism *in* jewelry to looking at the symbolism *of* jewelry. It seemed to me eventually that the market does not obviate meaning, and that the transaction of crafts or commodities leave tracings and these connect a tourist to a collector, an established artist to a less eminent imitator. Now I have come to believe that in the Southwest it is the real and spoken destinies and destinations of materials and jewelry that are always of greatest interest.

Acknowledgments

The author thanks the Heard Museum, Phoenix, and the Smithsonian Institution, Washington, for permission to use archival material. Marcus Amerman very generously granted permission to use an image of his work. Thanks are also due to Jonathan King and Jonathan Batkin who made comments on earlier drafts of the paper, and the Victoria and Albert Museum Research Department who offered a short Visiting Fellowship to pursue some of these issues.

Notes

1 "Indian arts and crafts" and "Indian jewelry" will be used to denote the historic use, "Native American crafts," as a descriptor.

2 Authoritative sources for the history and development of Native American jewelry include Adair (1989 [1944]); Woodward (1971 [1938]); Bedinger (1973); Frank and Holbrook (1990); Baxter (2001).

3 Silversmithing is the term used to describe the range of Navajo techniques used to make silver jewelry even though this may not conform to a more conventional definition.

4 Turquoise is a secondary mineral and a by-product of gold, silver, and copper mining. In the 1860s and 1870s gold, silver, and copper were found in New Mexico and Southeastern Arizona (Lamar 2000: 123, 133, 363, 402). Falkenstein-Doyle (2011) notes that by the 1890s turquoise mines, such as Cerrillos, had passed from Native American hands into American ones and that turquoise was purchased from mine owners.

5 Venables (1995: 20) argues that coin silver was the main source of silver for mainstream American silversmithing until the 1860s. Initially then, Native and mainstream American silversmithing used the same sources.

6 In Navajo and Pueblo culture, turquoise is associated with rain and for Pueblo culture, more particularly, fertility. For Pueblo communities a common ideational complex can be said to prevail, meaning that cosmology, topography, clan, season, and rituals are expressed through colors themselves linked to the sacred trinity of corn, beans, and squash (Whiteley 2005: 150–5; 2009).

7 This photograph is credited to both A. Frank Randall and the more prolific George Ben Wittick, both pioneer photographers who set up studios in the Southwest (see Fleming and Luskey 1992). Wittick's Navajo photographs include field studies and posed portraits, including weavers and silversmiths (see Broder 1990).

8 This photograph is credited to John K Hillers, who worked for both the US Geological Survey and the Bureau of Ethnology as a photographer (Fleming and Luskey, 1992: 108–10, 139, 178).

9 As the jeweler Gail Bird (Santo Domingo/Laguna) recently remarked (2009), turquoise cannot be skillfully drilled by balancing a pump drill in one hand on one's knee.

10 As the gem expert working for Tiffany & Co., and later its vice president, George Frederick Kunz encouraged the use of non-precious and colored stones in Tiffany's jewelry (Markowitz 2009). He was an early commentator on turquoise, and its importance to Native American culture, drawing from archaeological excavations and other testimony (see Kunz 1890: Chapter 2; 1913; 1915: 309, 352–3).

11 Some of this contrast is evident in Holder, who wrote of a Zuni lapidary as follows: "he sits in front of his door, and with his boring-tool, made by himself from the ancient type, bores a small but perfect hole in each bead" (1906: 757).

12 The photographs reflect different intents: one commercial portraiture, the other ethnographic survey through captioning. Randall and Wittick use the caption of "Navajo Silver Smith" (Bedinger 1973: 19), whereas Hillers notes "showing manner of …" (Fleming and Luskey 1992: 139).

13 The concept of "ordained" is borrowed from Chiappara (1997).

14 Fred Harvey Company Collection, Heard Museum, Phoenix, Arizona (hereafter referred to as FHCC), RC 39 (1) 1:9 Herman Schweizer to J.H. Macmillan, 09/18/1931.

15 Southwest traders were already organized into the United Indian Traders Association (UITA) from 1931. UITA's mandate included improving their own business practices and selling genuine Indian arts and crafts, as well as contributing to the welfare of Native communities. Batkin notes that UITA likely influenced the FTC's decision (Batkin 2008: 183).

16 FHCC, RC 39 (1) 1:10, Letter from Herman Schweizer to J.F. Huckel, 09/15/1932.

17 FHCC, RC 39 (1) 1:11, *United States of America Before Federal Trade Commission—In the Matter of Maisel Trading Post. Docket Number 2037. Finding as to the Facts and Conclusions. 08/21/1933;* FHCC, RC 39 (1) 1:12, *In the United States Circuit*

Court of Appeals for the Tenth Circuit September Term, 1934, No. 976, Federal Trade Commission, Petitioner v. Maisel Trading Post, Inc. Respondent.

18　FHCC, RC 39 (1) 1:11, *United States of America Before Federal Trade Commission—In the Matter of Maisel Trading Post. Docket Number 2037. Finding as to the Facts and Conclusions. 08/21/1933,* paragraph 10, p. 5.

19　FHCC, RC 39 (1) 1:11, *United States of America Before Federal Trade Commission—In the Matter of Maisel Trading Post. Docket Number 2037. Finding as to the Facts and Conclusions. 08/21/1933,* paragraph 3, p. 3.

20　Ibid., paragraph 6, p. 4.

21　FHCC, RC 39 (1) 1:12, *Federal Trade Commission. Docket No.2037, Maisel Trading Post, Inc. Brief in Support of the Complaint,* 1934 (?) p. 15.

22　FHCC, RC 39 (1) 1:11, *In the United States Circuit Court of Appeals for the Tenth Circuit September Term, 1933, No. 976, Federal Trade Commission, Petitioner v. Maisel Trading Post, Inc. Respondent. Brief for Respondent.* Statement of the Case, p. 6.

23　Ibid.

24　Ibid., p. 7.

25　Ibid., p. 8.

26　Ibid., p11.

27　FHCC, RC 39 (1) 1:11, *Statement of Facts Re: Indian Design Coin Silver Jewelry Manufactured by The H.H. Tammen Company, Denver, Colo.* 07/21/1933, signed by Carl Litzenberger, president.

28　Ibid. Tammen was not prosecuted because of its labeling (Batkin 2008: 183), although as is seen through the correspondence, many retailers were not fastidious in distinguishing "Indian design" and "Indian made."

29　FHCC, RC 39 (1) 1:11, Letter from Carl Litzenberger to H. Schweizer, 07/05/1933.

30　For further reading on the Fred Harvey Company refer to Howard and Pardue (1995), Weigle and Babcock (1996).

31　FHCC, RC 39 (1) 1:10, Letter from J.F. Huckel to H. Schweizer, 09/15/1932.

32　Ibid.

33　Ibid.

34　FHCC, RC 39 (1) 1:10, Letter from H. Schweizer to J.F. Huckel, 09/15/1932.

35　FHCC, RC 39 (1) 1:10, Letter from H. Schweizer to J.F. Huckel, 09/29/1932.

36　Ibid.

37　Ibid.

38　FHCC, RC 39 (1) 1:9, Letter from J.H. Macmillan Spanish and Indian Trading company, to H. Schweizer, 09/17/1931.

39　FHCC, RC 39 (1) 1:10, Letter from H. Schweizer to Carl Litzenberger, 07/26/1933.

40　FHCC, RC 39 (1) 1:11, *United States of America Before Federal Trade Commission, 08/21/1933; Docket Number 2037. Order to Cease and Desist in the Matter of Maisel Trading Post, Inc.*

41　FHCC, RC 39 (1) 1:11, *United States of America Before Federal Trade Commission—In the Matter of Maisel Trading Post. Docket Number 2037. Finding as to the Facts and Conclusions. 08/21/1933,* paragraph 4, p. 2.

42　Ibid.

43　René D'Harnoncourt, general manager, Indian Arts and Crafts Board, *Activities of the Indian Arts and Crafts Board since its Organization in 1936,* United States Department of the Interior, #173899, Victoria and Albert Museum National Art Library, General Collection, Box III.203R.

44　FHCC, RC 39 (1) 1:14, *United States Department for the Interior: Indian Arts and Crafts Board, Washington, Navajo, Pueblo and Hopi Silver.* 03/09/1937.

45　Ibid., p. 2. *Standards for Navajo, Pueblo and Hopi Silver and Turquoise Products.* Ibid., p. 2.

46　IHCC RC 39 (1) 1:12, *In the United States Circuit Court of Appeals for the Tenth Circuit September Term, 1934, No. 976, Federal Trade Commission,*

Petitioner v. Maisel Trading Post, Inc. Respondent, p. 23.

47 War restrictions meant that silver was no longer available, to the extent that manufacturing of silver goods almost entirely stopped (Venables 1995: 300). Native American smiths still continued to have access to silver, though some of this may have been coinage (Neuman 1971 [1943]).

48 IHCC RC 39 (1) 1:12, *In the United States Circuit Court of Appeals for the Tenth Circuit September Term, 1934, No. 976, Federal Trade Commission, Petitioner v. Maisel Trading Post, Inc. Respondent.*

References

Adair, John. 1989 [1944]. *The Navajo and Pueblo Silversmiths.* Norman, OK: University of Oklahoma.

Adamson, Glenn. 2007. *Thinking through Craft.* Oxford: Berg.

Appadurai, Arjun. 1988. "Putting Hierarchy in Its Place." *Cultural Anthropology,* 3(1): 36–49.

Bailey, Garrick and Bailey, Roberta Glenn. 1982. *A History of the Navajo: The Reservation Years.* Santa Fe, NM: School of American Research Press.

Barthes, Roland. 1972. *Mythologies.* Paladin: London.

Barthes, Roland. 1984. *Camera Lucida.* Flamingo: London.

Batkin, Jonathan. 2008. *The Native American Curio Trade in New Mexico.* Santa Fe, NM: Wheelwright Museum of the American Indian.

Baxter, Paula A. 2001. *Southwest Silver Jewelry.* Atglen, PA: Schiffer Publishing.

Bedinger, Margery. 1973. *Indian Silver: Navajo and Pueblo Jewelers.* Albuquerque, NM: University of New Mexico Press.

Bird, Gail. 2009. Remarks made at the British Museum conference, *Turquoise, Henry Christy and European Collections,* December 11–13.

Bird, Gail and Johnson, Yazzie. 2011. Manuscript of lecture given at the British Museum conference, *Turquoise, Henry Christy and European Collections,* December 11–13, 2009.

Broder, Patrician Janis. 1990. *Shadows on Glass: the Indian World of Ben Wittick.* Philadelphia, PA: Rowman and Littlefield.

Bsumek, Erika Marie. 2008. *Indian-made: Navajo Culture in the Marketplace, 1868–1940.* Lawrence, KS: University of Kansas Press.

Chiappara, Michael J. 1997. "Affirmed Objects in Affirmed Places: History, Geographic Sentiment and a Region's Crafts." *Journal of Design History,* 10(4): 399–415.

Conroy, Kathleen. 1975. *What You Should Know about Authentic Indian Jewelry.* Denver, CO: Pro-Grob Publishing.

Dilworth, Leah. 1996. *Imagining Indians in the Southwest: Persistent Visions of a Primitive Past.* Washington, DC: Smithsonian Institution Press.

Falkenstein-Doyle, Cheri. 2011. "Eighteenth-century Economy, Twentieth-century Merchandising: The Market for Turquoise in the Southwest, 1900–1940." Manuscript of lecture given at the British Museum conference, *Turquoise, Henry Christy and European Collections,* December 11–13, 2009.

Fergusson, Erna. 1936/7. "Crusade from Santa Fé." *The North American Review,* 242(2): 376–87.

Fleming, Paula Richardson and Luskey, Judith (eds). 1992. *The North American Indian in Early Photographs.* London: Phaidon.

Franciscan Fathers. 1968. *An Ethnologic Dictionary of the Navajo Language.* Reprint. St Michael's, AZ: St Michael's Press.

Frank, Larry, with Holbrook, Millard J. 1990. *Indian Silver Jewelry of the Southwest 1868–1930.* West Chester, PA: Schiffer Publishing Ltd.

Gilster, Elizabeth. 1999. "A Typology of the American Indian Silver Jewelry Industry." *Journal of the Southwest,* 38(4): 475–91.

Goring, Elizabeth. 2006. "Jewelry and Communication: Breaking the Code," juried Exhibition in Print. *Metalsmith,* 26(4): 6–9.

Herzog, Melanie. 1996. "Aesthetics and Meanings: The Arts and Crafts Movement and the Revival of American Indian Basketry." In B. Denker (ed.), *The Substance of Style: Perspectives on the American Arts and Crafts Movement*. Winterthur, DE: Henry Francis du Pont Winterthur Museum.

Holder, Charles Frederick. 1906. "Some Queer Laborers—Where Peaceful Living Is Preferred to Money Making." *The Craftsman*, X(6): 752–62.

Howard, Kathleen and Pardue, Diana. 1995. *Inventing the Southwest: The Fred Harvey Company and Native American Art*. Flagstaff, AZ: Northland Publications.

Hutchinson, Elizabeth. 2009. "The Indian Craze: Primitivism, Modernism, and Transculturation in American Art, 1890–1915." Durham, NC: Duke University Press.

Jackson, Devon. 2010. "The Elephant in the Portal." *Santa Fean* (August/September): 92–7.

Jernigan, W.E. 1978. *Jewelry of the Prehistoric Southwest*. Santa Fe, NM: School of American Research.

Johnson, Tim (ed.). 1998. *Spirit Capture: Photographs from the National Museum of the American Indian*. Washington, DC: National Museum of the American Indian.

Keane, Web. 2001. "Money Is No Object: Materiality, Desire and Modernity in an Indonesian Society." In Fred Myers (ed.), *The Empire of Things: Regimes of Value and Material Culture*. Santa Fe, NM and Oxford: School of American Research Press.

Kluckhohn, Clyde, W.W. Hill, and Lucy Wales Kluckhohn. 1971. *Navaho Material Culture*. Cambridge, MA: The Belknap Press of Harvard University.

Kunz, George Frederick. 1890. *Gems and Precious Stones of North America*. New York: The Scientific Publishing Co.

Kunz, George Frederick. 1913. *The Curious Lore of Precious Stones*. J.B. Lippincott & Co.

Kunz, George Frederick. 1915. *The Magic of Jewels and Charms*. London and Philadelphia, PA: J.B. Lippincott & Co.

Lamar, Howard R. 2000. *The Far Southwest 1846–1912: A Territorial History*. Alburquerque, NM: University of New Mexico Press.

Markowitz, Yvonne, J. 2009. "The Power and Allure of Gems." In Jeannine Falino and Yvonne J. Markowitz (eds), *American Luxury Jewels from the House of Tiffany*. Suffolk: Antique Collectors Club.

Matthews, Washington. 1968 [1883]. *Navajo Silversmiths*. In 2nd Annual Report of the Bureau of American Ethnology for the Years 1880–1881, Washington, DC, pp. 167–78. Palmer Lake, CO: Filter Press.

McEwan, Colin, Middleton, Andrew, Cartwright, Caroline and Stacey, Rebecca. 2006. *Turquoise Mosaics from Mexico*. London: The British Museum Press.

McLerran, Jennifer. 2009. *A New Deal for Native Art: Indian Arts and Federal Policy 1933–1943*. Tucson, AZ: University of Arizona Press.

Moore, Emily. 2008. "The Silver Hand: Authenticating the Alaska Native Art, Craft and Body." *Journal of Modern Craft*, 1(2): 197–220.

Moore, Laura Jane. 2001. "Elle Meets the President: Weaving Navajo Culture and Commerce in the Southwestern Tourist Industry." *A Journal of Women Studies*, 22(1): 21–44.

Morphy, Howard. 1995. "Aboriginal Art in a Global Context." In Daniel Miller (ed.), *Worlds Apart: Modernity through the Prism of the Local*. London: Routledge.

Mullin, Molly. 1996. "The Patronage of Difference: Making 'Art' not 'Ethnology.'" In George E. Marcus and Fred R. Myers (eds), *The Traffic in Culture: Refiguring Art and Anthropology*. Berkeley, CA: University of California Press.

Myers, Fred R. (ed.). 2001. *The Empire of Things: Regimes of Value and Material Culture*. Santa Fe, NM: School of American Research Publishing.

Neuman, David L. 1971 [1943]. "Navajo Silversmithing Survives." In David L. Neuman and Betty T. Toulouse (eds), *Navajo Silversmithing*. Santa Fe, NM: Museum of New Mexico/*El Palacio*.

Neuman, David L. 1971 [1950]. "Modern Developments in Indian Jewelry." In David L. Neuman and Betty T. Toulouse (eds), *Navajo Silversmithing*. Santa Fe, NM: Museum of New Mexico/*El Palacio*.

New, Lloyd Kiva. 1975. "Introduction." In Kathleen Conroy, *What You Should Know about Authentic Indian Jewelry*. Denver, CO: Pro-Grob Publishing.

Ortis, Raymond. 1935. *Indian Art of the Southwest: An Exposition of Methods and Practices*. Published and distributed by Southwest Indian Fair.

Russell, Inez. 2009. "Quest for Highest Quality." *The Santa Fe New Mexican*, 21 August.

Sandweiss, Martha, A. 2002. *Print the Legend: Photography and the American West*. New Haven, CT/London: Yale University Press.

Schrader, Robert F. 1983. *The Indian Arts and Crafts Board*. Albuquerque, NM: University of New Mexico Press.

Steiner, Michael C. 1983. "Regionalism in the Great Depression." *Geographical Review*, 73(4): 430–46.

Venables, Charles L. 1995. *Silver in America 1840–1940: A Century of Splendor*. Dallas, TX: Museum of Art/Harry N. Abrams Inc. Publishers.

Wade, Edwin L. 1985. "The Ethnic Art Market in the American Southwest, 1880–1980." In George W Stocking (ed.), *Objects and Others: Essays on Museums and Material Culture*. Madison, WI: University of Wisconsin Press.

Weigand, Phil C. and Harbottle, G. 1993. "The Role of Turquoises in the Ancient Mesoamerican Trade Structure." In J.E. Ericson and T.G. Baugh, *American Southwest and Mesoamerica: Systems of Prehistory Exchange*. New York: Plenum Press.

Weigle, Martha. 1989. "From Desert to Disney World: The Santa Fe Railway and the Fred Harvey Company Display the Indian Southwest." *Journal of Anthropological Research*, 45(1): 115–37.

Weigle, Martha and Babcock, Barbara A. (eds). 1996. *The Great Southwest of the Fred Harvey Company and the Santa Fe Railway*. Phoenix, AZ: Heard Museum.

Wheat, Joe Ben. 1982. "Early Southwest Metalwork." In Louise Lincoln (ed.), *Southwest Indian Silver from the Doneghy Collection*. Austin, TX: University of Texas Press.

Whiteley, Peter. 2005. "The Southwest 'Painterly' Style and Its Context." In Kari Chalker, Lois S Dubin, and Peter M. Whiteley (eds), *Totems to Turquoise: Native North American Jewelry Arts of the Northwest and the Southwest*. New York: Harry N. Abrams Inc./American Museum of Natural History.

Whiteley, Peter. 2009. "*Tsorposiniqw patangsi* [turquoise and squash blossom]: Puebloan Oppositions of the *Longue Durée*." Lecture given at the British Museum conference, *Turquoise, Henry Christy and European Collections*, December 11–13.

Witherspoon, Gary. 1977. *Language and Art in the Navajo Universe*. Ann Arbor, MI: The University of Michigan Press.

Witherspoon, Gary. 1981. "Silver and Turquoise Jewelry in the Navajo World." In *White Metal Universe: Navajo Silver from the Fred Harvey Collection*. Phoenix, AZ: Heard Museum.

Woodward, Arthur. 1971 [1938]. *Navajo Silver: A Brief History of Navajo Silversmithing*. Flagstaff, AZ: Northland Press.

The Journal of Modern Craft

Volume 5—Issue 1
March 2012
pp. 93–98
DOI:
10.2752/174967812X13287914145550

Reprints available directly
from the publishers

Photocopying permitted by
licence only

Daughterboats

Gail McGarva

Gail McGarva originally worked in Theatre in Education before re-training as a sign language interpreter in 1993. Having spent many years living on boats, in 2004 she decided to train as a wooden boat builder in the English Dorset town of Lyme Regis. The building of traditional boats of historical resonance is now her specialist area.

Abstract
Providing an insight into the workings of a builder of traditional wooden boats, this paper explores the author's specialist area, which is the building of replicas or "daughterboats" to vessels of historical resonance. The daughterboats are an expression of living history, each with a unique story to tell about their communities and their shores.

Keywords: traditional boat building, traditional wooden boat builder, clinker, pilot gig, lerret, replica boats, heritage vessels/historic boats.

I am a traditional wooden boat builder, with a passion for open working boats (Figure 1). The traditional clinker construction of these craft results in the beautiful lines and an evident skeletal, sculptural form. Boats built in this way were designed to have a long and hard life; they are strong and robust, yet their forms are elegant. While traditional in design, they continue to play an important role in present-day communities.

My specialist area is the building of replicas or, as I prefer to call them, "daughterboats" of these heritage vessels. I see my work as a means of breathing life into a historic practice. I am not making museum pieces but expressions of living history, which encourage a sense of belonging and a connection to the sea. Open boats are the unsung heroes of our coastal heritage, and each has a story to tell.

Fig 1 Gail McGarva in her Shetland daughterboat, *Georgie McDonald*. Photograph: City and Guilds.

My training began in 2004, at the Boat Building Academy in Lyme Regis. I had lived on boats for many years, but it wasn't until I journeyed to the source of the Thames on an old Navy boat that I was compelled to place them at the very core of my life. At the time I was working as a sign language interpreter, and I was in no way disenchanted with this work; but the pull of the boats became stronger and stronger. On my return from my Thames journey I visited the Academy for the first time and, on entering that workshop, I felt I had come home.

I had read an article about the story of Willie Mouat, a boat builder on Unst (one of the Shetland Isles, and the most northerly isle of Britain). Willie was the last remaining boat builder on Unst, and had no one to pass his skills onto. He talked about his work conserving historic Shetland boats in the Boat Haven Museum on Unst and his deep appreciation of the oldest surviving example of the local craft, the Gardie Boat of 1882. I was moved by Willie's story and his beautiful descriptions of the boats and was inspired to build a replica of the Gardie Boat—my first daughterboat—with Willie as my mentor. I wrote to him with my idea, acknowledging my own status as an outsider to the Unst community. I would be guided by him as to whether he felt it fitting or not for me to build such a boat. He telephoned me with his response. "We," he said (meaning the community) "would be delighted."

I in turn was honored to be embarking on such a journey. I traveled to the far North of Scotland to meet Willie and the Gardie Boat and came away with a head full of stories, an armful of notebooks—filled with measurements, building tips, photographs and sketches—and a rucksack full of Shetland rosehead copper nails.

Week by week Willie and I spoke on the phone, with him guiding me through the process of the build. I quickly became bilingual in the language of boat building, in English and in Norn—a dialect of Norwegian used by Shetlanders in the past, and retained in the terminology of the craft. I learned beautiful words like *humlabaund*, *kaeb*, and *stammeron*. The humlabaund is a strap of leather or rope that is looped around the square-loomed Shetland oar. This, in conjunction with the kaeb, a block of hardwood which is locked into the gunwale of the boat, acts as a holding mechanism for the oar as opposed to the use of a rowlock. The stammeron is a cant-angled oak frame, notched and beveled to fit around the ledges of the clinker planking, fastened with copper rivets in both the bow and stern.

It felt only right that my daughterboat should be launched on the island of Unst, from the same slipway where the motherboat had been launched in 1882. Willie chose the day of the Island show. The people of the community carried her to the water, and Willie and I rowed her out to sea. She glided through the water with majestic ease.

Since the building of this first daughterboat I have gone on to build several more, each time based on motherboats with a powerful connection to their communities and their shores.

In 2006 I went to Ireland to build a replica of a late eighteenth-century Bantry Bay gig. This daughterboat has been adopted for use by the Atlantic Challenge Project, which seeks to connect groups of young people across the ocean, asking them to work together under sail and oar. Following the ethos of Kurt Hahn, the founder of the Outward Bound movement, the Atlantic Challenge Project seeks to give a sense of adventure and challenge to young people, empowering them to take positive risks (as opposed to the negative ones that they may experience in other aspects of their lives). The Bantry gig is 38 feet long and is powered by ten rowers and a complex rig; sailing it on the open sea is a wonderful collective endeavor.

In 2007–9 I built two daughterboats based on the Cornish pilot gig *Treffry*, of 1838 (see Figure 2). This pilot gig is one of the most elegant of open working boats, 32 feet long with a beam of 4 feet, 10 inches and elm planking of only five-sixteenths of an inch thick. Originally there were two types of gig: a heavier craft for the carrying of cargo, and a lighter gig that would have piloted a larger ship, guiding it through unfamiliar waters. Competition among the pilot gigs was fierce

Fig 2 The ribcage of McGarva's Cornish pilot gig, *Rebel*. Photograph: Kerry Maguire.

and speed was of the essence, as the first gig to reach the incoming vessel commanded the fee to pilot the ship. In the revival of Cornish pilot-gig racing, speed is still of the essence and competition ever-strong. There were over 120 gigs on the start line of the 2011 championships on the Isles of Scilly.

In 2007 the late Ralph Bird, a master builder who had constructed twenty-nine gigs in his career, was on the brink of retirement. He offered to be my mentor and imparted to me all his patterns and templates. This gift felt very much like the passing on of the baton, from one generation of builder to the next. The renaissance of gigs and gig racing all along the South West coast and beyond is a testament to the passion people hold for these beautiful working boats.

Most recently I have built *Littlesea*, a daughterboat to the historic Dorset fishing vessel, the lerret of Chesil Beach, dating back to 1615 (Figure 3). It is difficult to wholly pinpoint the origins of the lerret. Some believe her to based on a Mediterranean boat called *The Lady of Lorretto*; others believe her influences to be Nordic. The intrinsic characteristics of this vessel have remained essentially unchanged for generations. An immensely beamy, flat-floored, double-ended vessel, it is designed to cope with the steep, stony terrain of the Chesil Bank. Eric McKee, a maritime historian who extensively studied the working boats of Britain, wrote with reference to the lerret:

> the combination of the steep beach,
> surf and cross current called for a boat
> with just the right amount of bearing or
> buoyancy in her ends. If her end was too
> full underwater, it would float before the

other end could be got clear of the beach. If this happened the current would swing the boat broadside on to the waves, which would swamp her. If, however, the end was not buoyant enough, waves would come in over the bows. The hull had to be beamy to support the weight of the nets and the crew, and this produced the rather pinched immersed sections and full topsides of this class of boat.[1]

My construction of the daughterboat was funded by the Queen Elizabeth Scholarship Trust. The aim was not only to preserve the lineage of the lerret but also to preserve the

Fig 3 *Littlesea*, daughterboat of a Chesil Beach lerret. Photograph: Matt Cotterill.

art of boat building by eye, working without the use of design drawings or construction plans. Roy Gollop, one of the few remaining Dorset boat builders with this skill, was my mentor for the project. Once more the skills from one generation of builder were passed to the next.

In their heyday of the 1800s, over 100 lerrets could be seen along the expanse of Chesil Beach but, in 2009, only a handful remained. *Vera*, built in 1923, was chosen as our motherboat, the guiding force behind our new-generation lerret. While I was building *Littlesea*, many people came forward with their memories, stories, photographs, and articles about the lerrets.

One man recalled helping his uncle and his lerret fishing crew in his teenage years of the 1960s. One night they pulled the vast "seine" net to the shore with an astounding catch of mackerel, weighing 992 stone. The net was bursting to the brim and it took until the small hours to load the catch into the willow pots and transport them across the Fleet, the lagoon that nestles behind the Chesil Bank. He was given a quarter share of £20 of fish for all his help in the landing of the fish and the lerret. This night heralded a rite of passage, for in recognition of all his hard work, he was from then onwards granted a man's full share.

The flavor of the fishing life of Chesil Beach is captured well in this bit of an interview I conducted with another local man:

> Shout out "mackerel spotted." Fish blown up, in goes the birds. That's another giveaway, if the fish are there, the birds are a giveaway. Sid would say "right, quick in the boat." Net ready to go, the crew would jump in quick, push off, travel up there … They would jump over the net like racehorses. So you used to say "pobble'em," not pebbles, but "pobble'em, pobble'em!" Grab some pebbles and throw the pebbles at the surface of the water to keep them down. Cause the fish will, if they hear a disturbance of water like that, they will dive down. Then haul in the net as quickly as you can.

Working with the Lyme Regis Museum, in the spring of 2011, we embarked upon a maritime heritage project, supported by the Heritage Lottery Fund, enabling us to preserve the memories of the lerret fisherfolk. This culminated in a touring exhibition along the Dorset coast in the summer of 2011 with the two lerrets *Littlesea* and *Vera* at the heart of it.

In the building of the daughterboats I have endeavored to be faithful to the essence of the original vessels whilst acknowledging that the new-generation boats will be used in a very different way to their precursors.

In traditional clinker construction, the edge of one plank overlaps the next, creating a stepped hull profile and accentuating the lines of the planking. If the lines are not sweet on the eye they shout out at you; if the lines are fair, then they sing. Clinker construction is a great test of skill. I love the demands it places upon you—there aren't really any hiding places.

I plank mainly in elm, and I adore the grain, the patterns, and the colors of this beautiful timber, as well as its unruly nature. When you are working it the wood loses its moisture and it wants to warp, cup, and change shape. It is incredibly playful, yet strong, durable and incredibly flexible. You need to get the plank locked in place as swiftly as you can and, once it is contained in the structure of the boat and is riveted, it settles.

In traditional clinker construction little has changed, as little has needed to. The craft has survived the tests of time. There is no use of glue or sealant between the overlapping planks. Everything relies on the accuracy of the planing, the precision of the ever-changing bevel along the length of each plank. If the bevel is good with no hard spots or shy areas, then the two mating surfaces (the "land" of the planks) will lie beautifully together. The hand-beaten copper rivets lock them in place.

Once the planking is completed the ribcage of the boat is steamed in place. It is an amazing process, massaging lengths of hot oak into the belly of the boat, creating a skeletal form of phenomenal strength and flexibility. It is almost as if it is a moving, living being, a whaleship.

This is what is most compelling about traditional wooden boats: people respond to them as if they have a life force of their own, even when on land but much more so on the water. They are drawn to these leviathan-like structures. Equally they are drawn to the stories that all boats have to tell about their communities and their waters. No shape of a boat evolved by chance; each is evocatively reflective of the nature of their shores and their role on the water, a perfect blend of form and function.

When building my daughterboats, I involve the local community in the whole process, from the inception of the project to the launching of the boat (Figure 4). Some

may help with the shaping and crafting of wood, some with the beating of a rivet. Some may come and join in the oiling of the boat; some may just come to watch the boat unfold, or stroke the elm and oak.

These daughterboats belong to their communities, not to any one individual. I believe it is a privilege to build such vessels. They demand patience and accuracy—and rightfully so. May these daughterboats speak out their stories in generations to come, as a testament to living history.

Note

1 Eric McKee, *The Mariner's Mirror*, 63(1) (1977): 42.

Fig 4 Launching the lerret, *Littlesea*. Photograph: Martin Haswell.

The Journal of Modern Craft

Volume 5—Issue 1
March 2012
pp. 99–102

DOI:
10.2752/174967812X13287914145596

Primary Text

Commentary

Glenn Adamson

Glenn Adamson is Head of Research at the Victoria and Albert Museum in London and is an Editor of *The Journal of Modern Craft*.

In the past decade the term "globalism" seems to have appeared everywhere, and craft studies are no exception. This journal has cast as broad a net as possible in geographical terms, and has often featured articles that contest ideas of authentic and discrete national traditions. Our effort has been paralleled in many other exhibitions, books, conferences, and magazines which stress craft's role as a mobile factor in cultural exchange. However, this instinct is by no means unprecedented. Five decades ago, Margaret Merwin Patch (1894–1988) undertook a globetrotting odyssey in which she examined the craft cultures of several continents. She began in July 1960, flying to Japan via Honolulu, and then carried on to mainland Asia, with stops in Hong Kong, Vietnam, Cambodia, Thailand, Indonesia, Singapore, Malaysia, Burma, India, Ceylon, Pakistan, and Afghanistan. She then carried on to the Middle East, visiting Iran, Lebanon, Syria, Jordan, Egypt, and Turkey and then—with a brief detour through Greece—went behind the Iron Curtain, traveling in Yugoslavia, Bulgaria, Romania, Hungary, Russia, Poland, and Czechoslovakia, before crossing back into Austria. She then continued in North Africa—Tunisia and Morocco—and after a brief respite back in the USA, carried on to Mexico and other locations in South America, finally finishing her journey in 1961. Everywhere she went, she interviewed local craft promoters, met artisans, viewed workshop practices, and took good notes.[1]

What would motivate a woman in her mid-sixties to undertake such an ambitious journey? As so often in matters of postwar craft history, the answer is Aileen Osborn Webb. Also then in her sixties, Webb was a close friend of Patch's, and was working toward an internationalized version of the

American Craftsmen's Council (as it was then called), which she had founded in 1943. Though it was Webb's vision that prompted the exploratory trip, Patch herself proposed the idea and also funded her travels personally. Following her return to the USA, she continued to act as an advisor to Webb on the formation and then the administration of the World Crafts Council. The two women also provided direct financial support for the WCC, initially running it (as Webb later said) "just by ourselves in a very amateurish way."[2] According to Liza Kirwin of the Archives of American Art, who collected Patch's papers in the 1980s and recorded an oral history with her, "The materialistic liberality of the two women certainly sustained each crafts organization, but it also created financial and to some extent conceptual dependence, which was unintended and regretted by Mrs. Webb and Mrs. Patch, as the latter makes clear in her interviews."[3]

Though she might have seemed a rather innocuous lady to others in the crafts scene (Museum of Contemporary Crafts director Paul Smith downplayed her importance in an interview I recently conducted), Patch was perfectly suited to the task of creating the first comparative international craft survey.[4] Born in Bloomington, Illinois, and educated at the University of Chicago Graduate School of Commerce and Administration (graduating in 1917) and Columbia University (in 1922), her primary profession was as an economic statistician. But she had also attended Cranbrook Academy in the 1930s in the company of her husband George Patch, studying with the modernist painters Zoltan Sepeshy and Wallace Mitchell. After the war she was

living both in Florida and in New York, where she befriended Webb, becoming a trustee of the ACC in 1953.

Given this formidable and varied background, it is perhaps less surprising that Patch felt confident in voyaging across the world on her own, or for that matter in assessing the diversity of craft production she encountered. The text printed here, previously unpublished but preserved as a typescript at the Archives of American Art, sees her contrasting the respect accorded to Japanese "artist-craftspeople" (in the American terminology of the time) with the lowly status of Indian artisans. It was a view shared by Webb, who said in a later oral history interview, "of the million weavers in India very, very few of them are creative. They're technicians; they're really artisans working in a factory only they're doing it in the home."[5] Patch viewed India as a great problem, because of its enormous population of often impoverished artisans, but also a great opportunity for cross-cultural collaboration. It was in India that she met Kamaladevi Chattopadhyay, founder of the All-India Handicraft Board and perhaps the closest analog to Webb working anywhere in the world, as well as Chattopadhyay's colleague Pupul Jayakar, and the craft historian D.N. Saraf, who organized her visits to workshops and museums. Saraf, interestingly, had just done something similar to Patch, visiting workshops in New Hampshire in the company of Allan Eaton to learn about American craft development strategies.[6] All of these figures' influence would be felt at the 1964 WCC conference in New York.[7]

In addition to the manuscript reprinted here, Patch produced a series of four articles which ran in the ACC's magazine *Craft Horizons* in 1962. Interested readers will want to seek out these texts, as they provide fascinating capsule studies of national craft economies at this date, as well as a record of her aesthetic impressions.[8] In Phnom Penh, Cambodia, she was spooked by the armed French troops filling the city but admired a loom in the National Museum, "one of the most beautiful I have ever seen … long and low, of dark wood inlaid with mother of pearl." In Java she enthused over the work of the Batik Institute, which combined preservation with innovative design, but was perplexed by stacks of batiks in the shop, "all folded up so that it is impossible to make a considered choice." In Egypt she recorded her visit to the famous tapestry workshop of Ramses Wissa Wassef, "one of [the] most moving experiences" of her travels. While she was there, a German television crew was too, and work was being packed up for an exhibition in Russia. And when she got to the Eastern Bloc herself, she was struck by the modernity of the crafts being produced under Communism. Czechoslovakia's government was doing a great deal for the arts—"far more than I saw anywhere else"—and at an exhibition of ceramics, textiles, and glass held in Red Square, Moscow, "the articles might have been from a contemporary design show anywhere, with a few exceptions which clearly showed their folk origin."

In the *Craft Horizons* travelog, as in the text reprinted here, Patch offers no easy answers about the dilemmas of craft development, and its countervailing impulses toward innovation and authenticity. Indeed she expressed doubt about the transformative potential of development projects involving industrial designers from abroad, then as now a commonly proposed solution. "The body of local workers, with their ingrained cultural patterns, will surely resist any such short-term efforts to make these changes," she wrote in her first *Craft Horizons* article. "If it is possible, or even desirable, to change the traditional craft designs of a country, it is more likely to be done by the buyer with a special order in hand, or by those living and working with local artisans who can over a period of time introduce new ideas and necessary standardization." As this passage suggests, Patch was no progressivist, but nor was she wedded to an ideal of intact traditionalism. What struck her most of all was "the extent of interpenetration of cultures and the isolation which fostered the development of such a fantastic variety of tongues; of the changing status of women and of the relation of men in the various caste and class systems; of the struggle everywhere for a better life, for independence of thought, and, not least, for recognition."

In all of her writings Patch shows herself to be a typical figure of the American postwar craft revival, when authorship and individuality were sacrosanct, and modernization was largely embraced as embodying democratic values. But she was also a forerunner of our own moment, when issues of global movement seem to us so much more palpable and ethically pressing. When the World Crafts Council was founded, its declared purpose was to recognize that "the crucial problem of a humanist culture is symbolized by the specific problem of the handcraftsman— that of finding his place in an industrialized international society."[9] It was a vision that

Patch had helped to formulate, and which still resonates today.

Notes

1 The Margaret Merwin Patch papers were acquired by the Archives of American Art in 1987. The collection comprises family records, correspondence, diaries, photographs, papers pertaining to the World Crafts Council, and an oral history recorded by Robert F. Brown in two sessions held in 1980 and 1984. See http://www.aaa.si.edu/collections/margaret-merwin-patch-papers-8327.

2 Aileen Osborn Webb, Oral History interview with Paul Cummings, May 7, 1970. Archives of American Art, Smithsonian Institution, p. 20.

3 Liza Kirwin, "Regional Reports," *Archives of American Art Journal*, 24(3) (1984): 26–39: 29.

4 Author's interview with Paul Smith, March 10, 2009.

5 Webb, Oral History interview, p. 24.

6 Patch, "Beginning of the World Crafts Council," typescript, 1982. Folder 12, Box 7, Margaret Merwin Patch Papers, Archives of American Art, Smithsonian Institution.

7 The proceedings of the first World Crafts Council conference were published in a limited run (copies exist at the Smithsonian, the ACC library, and Yale); for excerpts see Glenn Adamson, *The Craft Reader* (Oxford: Berg, 2010), pp. 199–205.

8 Margaret Patch, "ACC Forum: Craftsman's Odyssey," in four parts, published in *Craft Horizons*, 22(2) (Mar/Apr 1962): 50–3; *Craft Horizons*, 22(4) (July/Aug 1962): 52–3; *Craft Horizons*, 22(5) (Sept/Oct 1962): 52–3; *Craft Horizons* 22(6) (Nov/Dec 1962): 54–7.

9 Conference flyer, First World Congress of Craftsmen. Folder 1, Box 7, Margaret Merwin Patch Papers, Archives of American Art, Smithsonian Institution.

The Journal of Modern Craft

Volume 5—Issue 1
March 2012
pp. 103–108

DOI:
10.2752/174967812X13287914145631

Reprints available directly
from the publishers

Photocopying permitted by
licence only

Primary Text

The Craftsman

Margaret Merwin Patch

In Japan one finds the development and appreciation of the individual craftsman in a way that is almost unique in Asia. There are recognized artist-craftsmen, whose works are shown, admired, and for which sometimes enormous prices are paid. There are a number of associations which hold regular exhibitions of the work of the individual members, and galleries continually show the work of the individual ceramists or other artist-craftsmen. In speaking of one of the new buildings which are going up it is likely that the name of the craftsmen doing the mosaics or carvings will be mentioned along with that of the architect.

In India, on the other hand, this state of affairs is almost unknown. The recognized native artist-craftsmen whom I met or heard of could be numbered on the fingers of one hand. So far as I know there is no association of individual craftsmen, or regular exhibitions of their work. There were occasional exhibits of the work of individuals, and these were considered art events. However, such events are rare, and the situation does not seem likely to change under the present bureaucratic system.

Under such a system the organization is much like that of a modern factory, with the designers who design according to prescribed specifications, the executives who prescribe the requirements of a product, the supervisors who oversee the workers who supply the fingers and the know-how. It is of course different from the modern factory in that skill is more important than machinery. However, the traditional satisfaction of craft work, that of seeing the product through from beginning to end, is beginning to disappear in some cases as processes are divided for efficiency.

Where the country is large, as in India, the representatives of the bureaucracy multiply; there are regional deputy directors with complete staffs and design centers and field officers who travel continually, there are local centers for supervision, and finally at the bottom of this vast pyramid, the

Fig 1 Margaret M. Patch (left) and Aileen Osborn Webb. From *The First World Congress of Craftsmen* (New York: American Craftsmen's Council, 1965), p. 175.

craft worker. This is not to say that there are not specialists. Many areas are noted for distinctive products, and exceptional workers may be called on to head up workshops or to work in design centers. Under such a system the region may be well-known for certain distinctive products, and those acting as supervisors of design (with training often as painters or sculptors) may achieve some status. But so far as I could ascertain, there was no incentive or even opportunity for the individual to be creative, and no means for him to achieve recognition.

The status of the individual worker is really low in most of the Asian countries. In the smallest villages the craft worker works at his craft and also in the fields. In the larger villages there may be enough support for a greater specialization on [sic] the craft work and less outside work. In some cases a whole village may be given over to the production of one type of article. In most of these village or cottage industries the craft workers live on the same scale as their neighbors, which is low but not depressed. In the cities, however, and in some industries, there may be real exploitation of the workers. From Saigon to Fez I saw small girl-children at work knotting the carpets of the Orient, and embroiderers scarcely older working for a few cents a day. Weavers I saw working against time to produce yardage, their movements almost as mechanical as the machines which were gradually replacing them.

There are exceptions, of course. In the countries where the craft work is less highly organized the local craftsman may have his family workshop with a sales room attached, especially if it is in a tourist area. He is free to improvise and does so, usually within his basic traditional design, as his market develops. In some cases these local workers may achieve enough of a reputation to have his work sought out, but as a rule all such "folk artists" work in complete anonymity.

As in India there are occasional artist-craftsmen who have been trained in other countries, and they are outstanding in achievement and recognition. Schools of art are beginning to include crafts in their curriculum, but all too frequently the instruction is only in techniques and traditional design. I cannot remember meeting one outstanding graduate of such a school.

It is difficult to say exactly why this state of affairs exists in other Asian countries where [sic] there is such a flourishing and competitive art effort in Japan. In both cases the folk artists are still alive, even though they are all gradually succumbing to the machine. In both cases there have been periods of flowering of their respective cultures, with the architectural monuments and the ornamental objects developed during these periods still present, in situ or in museums or embedded in the traditional designs still produced. In both cases certain ceremonies and rituals still prescribe certain artifacts and these are still produced.

What then has made it possible for Japan to develop a reverence for her art objects and a respect for their creators which seems to be lacking elsewhere in Asia? Or conversely, what attitudes exist which curb

so sharply the appreciation in these other countries? One, of course, is the caste social system in India and the similar attitude in other countries which relegates any person doing physical labor to the bottom of the social scale. Another is the importation of the idea of the "fine" arts by the western European colonizers along with other aspects of their culture. This, it seems to me, is the most formidable obstacle.

It is the same in our country and in all of Europe. But whereas in Europe outstanding craftsmen have always been recognized, it has been a comparatively recent development to have organizations of such artist-craftsmen, to have museums and galleries open their doors to their work, and to have their work achieve the status of arts. It has been possible in our country and in Europe, and in Japan, because of the stimulating ferment throughout our culture, in which all of the arts participate. It has been made easier because there was no bureaucratic control which hampered free development, and because of the democratic attitude toward the arts.

In India and most of the other Asian countries the only artists recognized as such are the painters, who control the museums and galleries in the name of "fine" arts, and I would guess do not encourage craft participation. As crafts are at present under the control of a bureaucracy interested in them more from a social-economic point of view than that of art, this barring from participation in the art development of the country is not likely soon to be challenged.

My visit in India coincided with that of Queen Elizabeth. In city after city I saw the elaborate preparations made for her welcome, and could feel the excitement in the air. In New Delhi her visit, and mine, coincided with the celebration of Republic Day, the outstanding holiday of the year. For the week-long festivities people came from all over India, and because of the queen, from other commonwealth countries. In my hotel one could hear an amusing variety of British accents.

I had met Srimati (Mrs.) Kamaladevi Chattopadhyay, Chairman of the All-India Handicraft Board, in New York before I left, and had met her again at International House in Tokyo on her way home. She had shown great interest in Mrs. Webb's proposals for international cooperation, and when I arrived in India she did everything possible to make me feel welcome and to help me to see what I could of Indian handicrafts.

In the great Republic Day parade there were elephants with painted decorations and satin trappings, and tribesmen from the hills in primitive costumes and heavy silver jewelry. During the week there were programs of folk dancing by the tribesmen, and concerts of Indian music. There were many receptions. The queen inspected the Central Cottage Industries Emporium, the outlet for crafts from all over India, and was presented by Mme. Chattopadhyay with a handwoven sari of the sheerest gauze.

Mme. Chattopadhyay is the head of a vast and complicated organization which included not only the All-India Handicraft Board, but also the Cottage Industries and the Cooperative Union. It reaches into every area and almost every village of India. Originally and primarily a social welfare project, to save from starvation the millions of handicraft workers whose livelihood was threatened by machine production, the emphasis is still on the preservation of

traditional designs and techniques. Recently, however, an effort has been made to improve and modernize design, directing it toward tourist purchases and the export market.

At Mme. Chattopadhyay's request my craft-seeing program for India was handled by her Director of Handicrafts, Mr. D.N. Saraf. He had also only recently returned from the US, where he had attended a marketing seminar at Harvard, and had visited some of the New Hampshire workshops with Mr. Allan Eaton. He arranged an elaborate program for me, only part of which I was able to carry out. In each of the large cities there is a regional "deputy director", usually a design center, and a considerable staff, including field officers who go out to the villages.

An enumeration of the points of interest covered by my program in Madras will indicate not only their thoroughness in preparing it for me, but the range of craft activities in that one city. We went first to a trade fair, with temporary modern buildings exhibiting equipment of every kind, crowded with visitors. The craft building was large and well-designed and the interior beautifully arranged. I was most impressed …

The other experiment, by Mr. P. Sen, director of the Calcutta Design Center, was unusual and interesting as a human document. He had been distressed by the deterioration of the work of the metal workers in an outlying village, as compared with samples of the work in the same village fifty years ago. Feeling that perhaps the workers were completely out of touch with modern life, he brought two of the best into the city. For as much as a month he let them wander around, riding on streetcars and buses, going to movies, seeing the big buildings, going to the docks to see the ships unload and trying to get the feel of the city. Then he brought them back to the workshop at the center, and gave them metal and equipment. Their first products tended to be cramped and small, but as they were assured that they did not have to conserve the metal, they became bolder, and when I was there were turning out some very handsome pieces, wholly unlike anything they had ever done before.

Such individual flowering I would consider almost unique. Native designer-craftsmen in our sense are practically non-existent, due to several fundamental reasons:

1.   In the Indian culture the status of the hand worker is still very low. Even the best craftsmen receive a very low wage. I heard of no instance of a high price being paid for craft articles such as that which may be paid in Japan for a tea ceremony bowl by a name artist.

2.   This attitude toward the craftsman was compounded by the importation in colonial times of the western European cultural bias in favor of the so-called fine arts. Young persons are willing to study painting and sculpture and turn their training to designing for the crafts, who would not be willing to do the craft work themselves.

3.   Only the techniques of crafts, with completely stereotyped design, were taught in the two schools I visited, though in one the instruction in painting and sculpture was remarkably contemporary and free.

4.   So far as I know there is no encouragement or incentive for developing as a name artist in any one of the crafts, i.e., no competitions or exhibitions in which individuals as such may participate.

5.   India can still cling to those craft works of infinitely fine and microscopic detail called for by her culture only so long as there is a supply of cheap labor. It would be an economic impossibility to produce such work in any other culture. Similarly such works can only be produced even with low-cost labor only when the design is so traditional as to be almost automatic; any deviation would take too much time and add too much to the cost.

This last point is one which will limit the acceptance of many of the Indian craft techniques into the body of accepted international artist-craftsmen practices.

The Journal of Modern Craft

Volume 5—Issue 1
March 2012
pp. 109–114

DOI:
10.2752/174967812X13287914145677

Material Girls: Contemporary Black Women Artists

Reginald F. Lewis Museum of Maryland African American
History and Culture, Baltimore, February 12 to October 16,
2011.

Curated by Michelle Joan Wilkinson. Exhibition design by
Michelle Joan Wilkinson, Dave Ferraro, and Dawn Bennett.

Reviewed by Adrienne L. Childs

Adrienne L. Childs is an independent scholar.

Media is the message of "Material Girls: Contemporary Black
Women Artists," which features eight black female artists
who explore the expressive qualities of used tires, combs,
pink plastic bags, toy cars, human hair, and more (Figure 1).
Mounted by the Reginald F. Lewis Museum in Baltimore,
Maryland and curated by Michelle Joan Wilkinson, the group
exhibition highlights the traditional and non-traditional media
employed by Chakaia Booker, Sonya Clark, Torkwase Dyson,
Maya Freelon Asante, Maren Hassinger, Martha Jackson Jarvis,
Joyce J. Scott, and Reneé Stout. "Material Girls" is a refreshing
departure from group exhibitions of African American
artists based largely on their racial designation, with little or
no additional binding or critical structure. Moving beyond
the survey, the exhibition aims to examine these artists'
engagement with materiality as an outgrowth of the tradition
of black female creative production both in the context
of African American and African diasporan culture. "Given
black women's traditions with craft and fiber arts—consider
basket-weaving and quilting", states Wilkinson, "the skilled
fabrication of functional and aesthetic objects has long been
within their domain."[1]

"Material Girls" is organized by artist, each represented
by four to six pieces. Upon entering the exhibition space

Fig 1 Installation shot, *Material Girls: Contemporary Black Women Artists*, Reginald F. Lewis Museum of Maryland African American History and Culture, February 12 to October 16, 2011.

the viewer encounters works by Maren Hassinger. This section is dominated by Hassinger's evocative installation *Love* (2008), comprised of delicate pink plastic bags in a narrow pyramidal formation attached to the intersection of two walls. Moving through the gallery we encounter the tissue paper composition *Gestation* (2010), by Maya Freelon Asante, the fragility of which resonates with Hassinger's plastic bag installation.

Perhaps the boldest statements are in the central spaces of the exhibition where Chakaia Booker's abstract refashioned tire compositions are a commanding force. The relatively small space afforded Booker's assertive sculptures, however, limits the viewer's ability to absorb works such as her

menacing, primordial abstraction *Black Hole* (2001). Similarly, the dramatic presence of Martha Jackson Jarvis's large-scale organic formations made of stones and glass would have benefited from more space (Figure 2). Sonya Clark's intriguing assemblages of combs and a wreath of human hair round out the first part of the exhibition.

Although not specifically articulated by the didactic wall text, the works featured in the first section are arguably of a more formalist character. Whether the craggy surfaces of Jackson Jarvis's stones, or the texture of Booker's tires, the largely abstract compositions exploit the distinctiveness of the material in creating an aesthetic statement. However, interpretive panels reveal how the works are also informed

Fig 2 Martha Jackson Jarvis, *Umbilicus*, 2006, volcanic stone, glass, wood, 23 × 76 × 20 ft (7 × 23.2 × 6 m). Photograph: John Woo.

by cultural, social, and political concepts. For instance, Clark's *Plain Weave* (2008), an assemblage of plastic combs and thread, references traditional West African kente

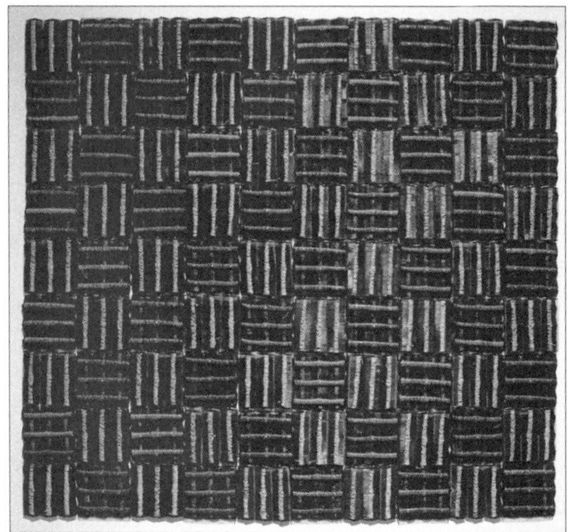

Fig 3 Sonya Clark, *Plain Weave*, 2008, Plastic combs, thread, 45 × 50 in. (114.3 × 127 cm). Photograph: Amy Weiks.

cloth as an evocation of cultural memory (Figure 3).

Torkwase Dyson, Joyce J. Scott, and Renée Stout are featured in the last part of the exhibition space. These women fashion materials in a narrative and figurative manner, often drawing upon African spiritual practice and ritual material culture. While Jackson Jarvis's totemic abstraction *Crossroads/ Trickster I, II, III* (2006) conceptually suggests the Yoruba guardian of the crossroads, Eshu Elegba, Dyson, Scott, and Stout take a decidedly figurative approach in their various modes of expression. The importance of African cultural production as a source—a critical thematic thread of this exhibition—is embodied in Scott's *Inkisi: St John the Conqueror* (2009; Figure 4). This assemblage of glass bottles, beads, wire, and coral responds to the corporeal structure and ideology of the *nkisi*, a ritual object from the Central African Kongo peoples. The

Fig 4 Joyce J. Scott, *Inkisi: St John the Conqueror*, 2009, collected glass bottles, glass beads, wire, thread, coral, 17 × 24½ × 17 in. (43.2 × 62.2 × 43.2 cm.). Photograph: Michael Kyorta.

traditional *nkisi* is also an amalgam of various materials such as wood, glass, metal, clay, shells, cloth, and herbs, and acts as a medium through which the powers of the dead are made available to the living. A container for spiritual forces, the *nkisi* has medicinal qualities and harnesses the energy of the spiritual world.[2] Scott's *Inkisi*, a doll-like figure shrouded with small bottles filled with red coral "roots," invokes the Kongo *inkisi* as power object, healer, and container of spirits, while maintaining the whimsical decorative quality of the beads and coral embellishment. The provocative tension between the political and the ornate characterizes Scott's dazzling beadwork.

"Material Girls" nests the production of the eight black female artists within two general ideological structures. The core interpretation of the works exhibited appears to revolve around what the curator calls black women's "adaptive strategy of 'making it work' into making art work."[3] Many of the artists were influenced by generations of women in their families who practiced traditional domestic crafts, particularly sewing. The exhibition demonstrates how the improvisational and experimental nature of black women's domestic artistry spawned the creative energies of subsequent generations. This notion is in turn presented as part of a larger traditional African practice of transforming everyday materials into powerful, energized objects. In her foreword, Lowery Stokes Simms, curator at the Museum of Arts and Design in New York, states that the transformation of "mundane materials is so prevalent in the work of African creators globally, that it could almost be identified as an ethnically determined aesthetic."[4] Simms's statement situates the works within a trajectory between African traditions, African American cultural production, and global contemporary African diasporan modes of expression.

While theories of African diasporic aesthetics and the power of traditional folkways indeed provide a critical foundation, I wondered as I viewed the exhibition, what is lost when these works are circumscribed by ethnic or racial themes. For instance, would the discussion have been strengthened by a historically grounded exploration of the links between this group of artists and the fiber art revolution of the 1960s and 1970s? Notably, African American artists Faith Ringgold and Barbara Chase-Riboud, both mentioned in the exhibition catalog as progenitors of this contemporary group, were part of the major craft revival of the 1960s and active in what Elissa Auther calls the "feminist recuperation of women's craft traditions …"[5] They were among the women that helped to break the hierarchical barriers between art and craft, legitimizing the use of alternative materials and through them asserting the voice of the female artist in the American art world. Indeed, there is a distinctly feminist bent to the work of "Material Girls" Asante, Scott, Stout, and Clark, who interrogate female identity, sexuality, beauty, motherhood, and more through their boundary-bending

materials. To abstract such artists from historical narratives marginalizes them as somehow disconnected from the shifts in the larger landscape of American art. Although feminist ideologies and the art/craft divide do not surface in the kind of traditional representation Simms posits as foundational in African diasporan expression, I would submit that the contemporary female artists in "Material Girls" are acutely aware of the boundaries they challenge when working in a "craft-like" mode.

The need to focus on and highlight the complex histories of African American production is a by-product of the unfortunate fact that black artists have been traditionally excluded from mainstream canonical histories. "Material Girls" evocatively takes on the complex challenge of expanding how we understand and evaluate the influences and concerns of African American artists. "Material Girls: Contemporary Black Women Artists" makes a vital contribution to the ongoing efforts to expand the depth and breadth of how we construct and present the innovation, skill, and creativity of African American artists and in so doing reveals a largely overlooked chapter in the history of American craft.

Catalog

Wilkinson, Michelle Joan (ed.). *Material Girls: Contemporary Black Women Artists*. Baltimore, MD: Reginald F. Lewis Museum of Maryland African American History and Culture, 2011.

Notes

1 Michelle Joan Wilkinson (ed.), *Material Girls: Contemporary Black Women Artists* (Baltimore, MD: Reginald F. Lewis Museum of Maryland African American History & Culture, 2011), p. 11.

2 Wyatt MacGaffey, "Complexity, Astonishment and Power: The Visual Vocabulary of Kongo Minkisi," *Journal of Southern African Studies*, 14(2) (1988): 188–92.

3 Wilkinson, *Material Girls*, p. 16.

4 Wilkinson, *Material Girls*, p. 8.

5 Elissa Auther, *String Felt Thread: The Hierarchy of Art and Craft in American Art* (Minneapolis, MN: University of Minnesota Press, 2010), p. 25.

The Journal of Modern Craft

Volume 5—Issue 1
March 2012
pp. 115–118

DOI:
10.2752/174967812X13287914145712

Dark Matter: Art and Politics in the Age of Enterprise Culture
Gregory Sholette

London: Pluto Press, 2010. 256 pages, 41 images, notes, bibliography, appendix (artists' groups survey 2008), index. US$30.00/GBP16.00 (paperback). ISBN: 9780745327525

Reviewed by Dave Beech

Dave Beech is an artist in the collective Freee, teaches at Chelsea College of Art, London, is Vice-chair of ixia, and is currently writing a book, *Art and Value*, that provides a Marxist economic analysis of art.

Gregory Sholette's book *Dark Matter* is an important and timely contribution to two essential projects for the Left, and radicals generally, in contemporary art and art history, perhaps even culture full stop. The central contention of the book is that there is a lot more to the production of art and the production of culture than is evident in the commercial art world. This extends the territory for thinking about the relationship between art and all that creative activity that is not authorized as art. First, it documents a sector of critical art practice that has gone largely unrecognized in the mainstream coverage of art since the 1960s, namely, the explicitly political art of self-organized artists on the fringes of the art world in the late 1970s and early 1980s. And second, it develops a theory of understanding this art via some of the most intelligent political thought around, in particular theories of Tactical Media, post-Situationist cultural techniques, precarity, and the sociology of economics. The theory is vital for repositioning the apparently "failed" works that he retrieves from the recent past, but it is the detail and breadth of the examples he gives, that is, the empirical part of the book, which is the real strength of this publication. And although emphatically not about craft as such, this book will have much relevance to the vanguard craftivists of recent years that extend the paradigm of the vital, yet marginalized, art practices that Sholette sets out to recoup.

The title of the book refers to the absence of large sections of artistic practice from the history, debate, institutions, and markets for art. It includes, Sholette tells us, "makeshift, amateur, informal, unofficial, autonomous, activist, non-institutional, self-organized practices – all work made and circulated in the shadows of the formal art world" (p. 1). This means putting the emphasis on cultural production in the broadest sense as being based on social values, not the economic values placed on objects in the art market. If we take art to include this enormous body of invisible, or barely visible, practices, we inevitably confront a much bigger entity than art is usually taken to be. Also, and more importantly, this "seemingly superfluous majority" requires us to transform radically our conception of what art is. A great many theories of art over the past hundred years or so have emphasized the incorporation of art by market forces, with concepts such as the culture industry, commodification, recuperation, and post-Fordism occupying critical territory in relation to that immersion of art in the economy, and concepts such as the creative economy, flexible organization, and urban regeneration reflecting the potential for art to work more productively within the capitalist economy.

The concept of "dark matter," in my reading, challenges the very basis of the argument that the majority of art, or all the art that really matters, is always already up to its neck in capital. Metaphors like dark matter, Sholette says, "are at best a means of

visualizing that which cannot be seen" (p. 45) but this underplays the significance of a concept that seems to combine the qualities of Marx's formulation of the "proletariat" with Rancière's *part des sans part* and Hardt and Negri's "multitude" applied specifically to contemporary culture. It carries with it the political aspirations of anti-art, the strategic goals of the avant-garde, and the basis of a new sociology of art, as well as the kernel of a critique of the standard economics of art. Dark matter is an extremely suggestive concept for those of us who oppose art's dominance by the art market, the economics of super-galleries, the colonization of art museums by corporate sponsors and the tourist industry, and so on. Sholette, however, does not overstate the implications of the concept of dark matter. In fact, he is rather reserved about the claims he makes for it. The "isolated flashes of defiance" of dark matter "are at best disjointed acts of insubordination. They do not necessarily knit together as sustained politics, and they are not inherently progressive or democratic" (188).

It is true that the dark matter documented in Sholette's book does not add up to a sustained politics, but it does make for a lucid narrative. Sholette does a brilliant job of bringing a vast range of material together, documenting, contextualizing, and theorizing it in a sensible order that knits divergent practices together into a largely coherent landscape. Not all the art groups he refers to are as dark or invisible as others. The Yes Men, Critical Art Ensemble, Chto Delat?, and Group Material are not exactly invisible, but their presence in this book is justified by the fact that they occupy the same politically committed territory as their invisible comrades, Paper Rad, The Waitresses, REPOhistory, The United Victorian Workers Union, and "Blue Funk n. Inf., chiefly Brit. A state of great terror." Sholette makes sense of this divergent collection of practices, partly through a kind of curatorial dexterity in which he assembles relationships between groups, and partly through the vivid and compelling depiction of the political world that they share. As a result, the book moves in two opposite directions, extending the knowledge of contemporary art beyond the gallery and tightening the relationship between the interior and exterior of art's institutions at the same time. Afterward, the reader has both a broader sense of the recent history of contemporary art and a stronger sense of how it is structured, what is at stake in it, and where its limits are.

The darkness of dark matter is a political kind of invisibility. Artists, works, and projects are invisible *vis-à-vis* power and wealth. Following Laclau and Mouffe's critique of the "privileged signifier" of class in Marxism, and Michel de Certeau's dissident tactic of the everyday, as well as Tactical Media's "indefinite politics," dark matter is hedged, mediated, and diverse. It is a field rather than a force. However, Sholette bases his politics on economics, which gives the concept more traction. To do this, he draws on David Harvey's Marxist analysis of neoliberal capitalism, in which the contrast between extreme wealth and abject poverty is shown to have been systematically exacerbated since the 1970s. He applies this directly to the politics of artistic visibility and invisibility, as a contrast between a minority of big "winners" and a mass of miserable "losers." This is linked to the mainstream economic

argument of "superstar economies" in which financial rewards are made on a "winner-takes-all" basis. Here Sholette's economics is flawed. Art is not a "winner-takes-all" economy. Artists, unlike producers in capitalist enterprises, are not motivated primarily by financial gain. What is more, one cannot say that economically successful artists "take all" of what economically unsuccessful artists accomplish and obtain (which are, largely, non-pecuniary benefits).

Sholette's flawed economic arguments don't stop there. Sholette also makes other problematic claims. "Consider the destabilizing impact on high art," he says, "were some of these hidden producers to cease or pause their activity" (p. 1). There is an argument to be developed here about the dependence of art, even the most conceptual, on the craft, design, and technical skills of others. Sholette claims "it is possible to imagine a thought experiment that would measure [dark matter's] aggregate impact on the art world – if say, one were to organize an art fabricators' strike, or a boycott of international art magazines, demanding these journals cover creative work made by the glut of artists who go unobserved in the art world" (p. 41). However, there is a fundamental problem with the idea of an "art strike." Art production is not wage labor. Normally when workers go on strike the capitalist or firm who pays their wages loses labor power and therefore profit. This is why striking is effective. It uses the latent economic power of labor against those who wield power over labor on a daily basis. However, no capitalist or firm loses out on labor power when artists go on strike. It is not an effective economic instrument of struggle.

"Dark matter" is not, in itself, the solution to the problematic divisions and hierarchies within cultural production, nor is it the last word on the tense relationship between culture and the social world. Asserting difference from the commercial art world might appear virtuous, but it is merely adapting to an objectionable situation of cultural division if it does not press that difference into a force for change. Nevertheless, for those of us trying to transform the current configuration of cultural and social division, Sholette's book is a precious document of attempts to address the central issues of art's wider social significance and a serious contribution to a debate that needs to be expanded.

The Journal of Modern Craft   Volume 5—Issue 1—March 2012, pp. 115–118

The Journal of Modern Craft

Volume 5—Issue 1
March 2012
pp. 119–122

DOI:
10.2752/174967812X13287914145758

Book Review

The Subversive Stitch: Embroidery and the Making of the Feminine
Rozsika Parker

London and New York: I.B. Tauris, 2010. xxii + 248 pages, 91 black and white images, notes, glossary, index. US$28.00/GBP14.99 (paperback). ISBN: 9781848852839

Reviewed by Eileen Boris

Eileen Boris is Hull Professor and Chair of the Department of Feminist Studies at the University of California, Santa Barbara, and specializes in women's labors in the home and other workplaces.

For more than a century, Western feminism has envisioned both a revaluing of femininity and the end of gender distinction, though too often difference and equality appeared in opposition, as if celebrating femaleness and demanding the same rights as men were contradictory goals. Rozsika Parker taps into this discussion by simultaneously deconstructing the deployment of embroidery to produce hegemonic womanhood and its appropriation by women for their own ends. "Limited to practicing art with needle and thread," she shows through a wealth of historical examples, "women have nevertheless sewn a subversive stitch—managed to make meanings of their own in the very medium intended to inculcate self-effacement" (p. 215).

Not that Parker uses cultural studies approaches to unravel the meanings of needlework constructed over the centuries, though speculation about mother-daughter bonds foreshadows her practice as a psychotherapist and recent scholarly interest in affect. Still, in method and argument, *The Subversive Stitch*, first published in 1984, is very much a product of its time: that moment when women's liberation still seemed possible as a collective rather than individual achievement, before the full weight of the neoliberal restructuring of the welfare state, with its defunding of the arts, and the religious and political right's assault on women's bodies. Since then, global capital has embraced the modernization of gender, including same-sex marriage and gender bending, but still relies on divisions of labor by sex, race, class, and nation. That *The Subversive Stitch* has held up so well in the quarter century since its first edition testifies to Parker's keen eye, broad knowledge, and sharp intellect.

Parker is at her best in unraveling stitches, explicating their symbolic and pictorial images. She understands embroidery as labor and places its practice in the social spaces of women's lives: the family, the school, and the church. Particularly impressive are the multiple levels on which the book operates. Parker has written an intellectual history of ideas about women's nature and social place; a labor history of the work process; and an art history of the iconography and techniques of tapestries, samplers, and copes. These modes of presentation work together when she links changes in gender ideology to larger social circumstances and shifts in representational motifs. For example, late eighteenth-century rural genre scenes reinforced "the very virtues demanded of bourgeois women—cheerful dutifulness and simple neatness" by depicting the poor as contented, domestic toilers, "concealing oppressive labour relations" (p. 143). At the same time, novelists like Samuel Richardson associated embroidery with female dependence, making it leisure rather than labor, equated "with feminine seductiveness" (p. 119).

This is no simple story of progress but rather a "spiral" of declension narratives. Respect as art workers in the medieval and Elizabethan periods, when men dominated the most prestigious sections of the craft, dissipates as embroidery becomes feminized and women domesticated. As early as the seventeenth century, "Working-class women were employed as sweated labour in trades associated with embroidery, and middle-class women became embroiderers because the craft's aristocratic and feminine associations made it an acceptable occupation" (p. 108). The division of art from craft in this first declension coincided with the association of the decorative with the feminine. "The eighteenth century … at least allowed that women achieved some satisfaction in embroidery," Parker asserts. "In the nineteenth century, unless embroidery was performed as a moral duty, in the spirit of selfless industry, it was regarded as sinful laziness—redolent of aristocratic decadence" (pp. 153–4).

Together, the Arts and Crafts and women's suffrage movements reversed such contempt. British suffragists deployed embroidery "to change ideas about women and femininity" (p. 197). As Parker perceptively explains, in their banners "Paint and embroidery, and masculine and feminine symbols, share the same space and make a political point—the demand for equality, not androgyny" (p. 199). This progressive use of embroidery fell into abeyance in the mid-twentieth century but reemerged with women's liberation. But today's embroidering artists, she contends in a new introduction, have lost the collective and revolutionary elan of thirty years ago.

Focused on contemporary artistic production, this introduction reads as an afterthought. Parker fails to draw upon new directions in Women's History. The agency that she finds in privileged women more recently scholars have extended to the lower orders. Her discussions of materials and styles slight the influence of Empire. Also missing is intersectional analysis with its focus on race and nation, though current researchers could learn something from her class analysis. Nonetheless, in championing the political over the universal, *The Subversive Stitch* underscores the power of pleasure to change society, not merely individuals. It stands as a lasting testimony to the brilliance of this pioneer in feminist art criticism. Rozsika Parker passed away in November 2010: what she wrought lives on.

The Journal of Modern Craft   Volume 5—Issue 1—March 2012, pp. 119–122

The Journal of Modern Craft

Volume 5—Issue 1
March 2012
pp. 123–126
DOI:
10.2752/174967812X13287914145794

Reprints available directly
from the publishers

Photocopying permitted by
licence only

Extra/Ordinary: Craft and Contemporary Art
Maria Elena Buszek (ed.)

Chapel Hill, NC: Duke University Press, 2011. 344 pages,
8 black and white/73 color images, notes, index.
US$89.95 (hardback); US$24.95 (paperback).
ISBN: 9780822347392 (hardback);
9780822347620 (paperback)

Reviewed by
Meredith Goldsmith

Meredith Goldsmith is formerly a Curatorial Associate at the
Contemporary Arts Museum, Houston and is currently in
the PhD Program in Visual Studies at University of California,
Irvine.

The last five years have witnessed a surge of publishing
addressing the proliferation of craft-identified materials and
practices—and not just in fine art. Knitting, embroidery,
gardening, and culinary arts seem to be having a renaissance
in mainstream culture. Add to that the phenomenon of
craft-activism, and the boundary-crossing condition of craft is
undeniable. Into this energized sphere comes the anthology
Extra/Ordinary: Craft and Contemporary Art, edited by Maria
Elena Buszek. Not a foundational reader, nor an academic
history, the volume is a kind of "the state of craft in art"
after postmodernism. The selection of essays introduces
myriad hybrid practices as well as the academics, writers,
and practitioners—some recent graduates, some established
professionals—currently working in this under-theorized
area. Rather than an attempt to narrow the field, this book
attests to an expansive, decentered, vibrant condition. Buszek
avoided having too heavy an editorial hand; each author's
voice is maintained rather than serving one unified narrative,
and the book is the better for it.

In some ways Buszek sees the prevalence of a "return" to handwork as a response to social and corporeal domination by twenty-first-century technologies, which diminish traditional relationships, materiality, and touch. Perhaps, she considers, we are witnessing a continuation of Arts and Crafts-era reactions against perceived socio-cultural losses to industrialization. On the other hand, unlike the Arts and Crafts Movement's ultimately rarified objects, or the subsequent adoption of art critic Clement Greenberg's intra-media critical model and Modernist ideals by studio craftspeople, or even postmodernism's celebration of the popular, the folk, and the marginalized, Buszek finds that the current embrace of the handmade comes at a time when crafting is more mainstream, profitable, and youthful than it has been in generations. Today craft is mobilized less often for identity politics and more in the service of an anti-globalization ideology. This places our newest generation of artists in a unique position. "These essays demonstrate that many artists drawing on craft culture do so in ways that revel in its boundary-crossing potential," she writes, "but in ironic and ambivalent ways different from (or even poking gentle fun at) the romantic or moralizing sensibility with which their predecessors often approached these media" (p. 12).

Extra/Ordinary is loosely divided into four sections. In "Redefining Craft: New Theory," contributors offer historical and theoretical frameworks for analyzing craft, positioning it as a structure through which 1960s art movements might be reconsidered. M. Anna Fariello suggests that craft would be more sufficiently historicized and theorized

if considered within a broader history of making—anything in material culture, rather than merely within art history—including industrialization and hands-on education. Paula Owen argues that process-oriented postmodern strategies of physical repetition and encounter are indebted to craft paradigms and can be theorized as such. The essays in section two, "Craft Show: In the Realm of 'Fine Arts,'" offer case studies of institutional art's appropriation of craft, and craft artists' facility with fine art paradigms such as conceptualism and self-interrogation. Karin E. Peterson's essay reminds us that long before "The Quilts of Gee's Bend" there was the 1971 exhibition of Jonathan Holstein and Gail van der Hoof's quilt collection at the Whitney Museum of American Art. She rigorously demonstrates that it was the cultivation of the "modern eye" through abstract painting that epistemically reframed quilts—but only certain kinds of quilts—as fine art to be hung on a wall. Elissa Auther does the impossible by offering a new reading of Andy Warhol. She argues that his decorative wallpaper rebuffed both Abstract Expressionism and dismissals of it, and shows how his embrace of the decorative's associations with effete homosexuality and femininity influenced later artists such as Robert Gober. The "Craftivism" section begins with Betsy Greer's account of coining that term, and continues with historical perspectives on the artists, events, and strategies of this newly self-identified movement. Finally, the fresh theories and analyses in section four, "New Functions, New Frontiers," suggest that it is not craft that needs to be re-identified for acceptance by art institutions or mainstream culture.

Rather, through this elusive, multivalent, ever-present thing we call "craft," we might change those very cultural dominants.

Especially compelling are the essays by Louise Mazanti, Kirsty Robertson, and Lacey Jane Roberts. In "Super-Objects: Craft as an Aesthetic Position," Mazanti proposes that craft is a means of making objects that are privileged aesthetically, but attached to the mundane. This means that craft can circumvent the historical avant-garde paradox that Peter Bürger famously theorized, in which an object cannot be both a detached fine art object and claim influence on everyday life. In pursuit of total aesthetic autonomy, fine art has often turned to material culture and craft, moving toward the everyday to achieve critical distance from the art world, and thus an avant-garde position. Yet craft has never had to overcome an art/life dichotomy, because it has always embodied both. Mazanti's discussion of the super-object goes beyond issues of craft materiality, and provides a rich theoretical approach, dovetailing material culture studies and avant-garde theory in a way that warrants further exploration.

Kirsty Robertson's essay "Rebellious Doilies and Subversive Stitches" proposes historicizing activism through craft production—think hand weaving during Indian Independence, or the AIDS Memorial Quilt. But rather than a one-dimensional touting of craft as an activist's universal remedy, Robertson argues that after earlier generations of vilified crafting, "it is precisely the *apparent* novelty of activist crafting that grants efficacy to actions that depend on the reversal of stereotype rather than links with second- and third-wave feminism …

Is it possible," she wonders, "that the way knitting, embroidery, and quilting are used to make political change in some spheres *requires* their subjugation in others?" (p. 186: my emphases). Robertson argues that the art-world success of artists such as Tracey Emin, the mainstreaming of textile crafts in an increasingly conservative culture, and the current activist investment in knitting, are all ultimately related to the evisceration of the textile industries in Western Europe and North America. Where formerly ubiquitous textile craft was "naturally" feminine, and therefore a site of broad feminist critique, it is now a more specific activity which can be used to take a political stance. Robertson's chief concern is that craftivists wake up to their position within a long continuity of protest crafters, especially women.

Also challenging essentializing identities of craft and crafters, Lacey Jane Roberts argues that critical craft theorists have much to learn from queer theory. "The tactics of reclamation, reappropriation, and disidentification used in queer theory and praxis give non-normative identities agency as well as question the seemingly stable systems that render them as other," she writes. "These tactics acknowledge stereotypes, transpose them, and then subvert them to form new models of identity" (p. 245). What can such tactics do for craft and crafters? Rather then trying to overcome its marginalization by art history, craft might speak from, or even amplify its shameful, grossly generalized qualities. Among Roberts' examples are Liz Collins' *Knitting Nation*—a "spectacle of slowness" and feminized patriotism perverted with sweatshop-like laboring—and Josh Faught's installation *Nobody Knows I'm a Lesbian*, in which the artist tells an obvious lie, yet presents it through "authentic" craft. Empowered through this kind of tricksterism, crafters can effectively critique the entire dominant culture, not just art historical discourse.

Suggesting tactics for more "effective" craft, though, prompts the question: who is the intended audience for *Extra/Ordinary*? Although it is entirely possible that a general crafter or hobbyist will pick up this book and glean something from it, it is more likely that its readers will be studio craftspeople, art world initiates, and art students just finding their aesthetic voice, defining their place in the art establishment. I agree with Buszek's assessment of the newest generation of fine artist-crafters and would add that they are also unique in the prevalence of "theory" in their education and professionalization as artists. Many essays in this book will reinforce that aspect of their exploration.

THE JOURNAL OF MODERN CRAFT
Notes for Contributors

Articles should be of a length between 5,000 and 6,000 words including notes and references, with 8 to 10 images. They should include a three-sentence biography of the author(s), an abstract and 5–8 keywords Exhibition and book reviews are normally 750 to 1,250 words in length.

The Journal of Modern Craft invites persons wishing to organize a special issue devoted to a single topic to submit a proposal comprising a 100-word description of the topic, together with a list of potential contributors and paper subjects. Proposals are accepted only after review by the journal editors and in-house editorial staff at Berg.

The Journal is published three times a year, in March, July and November. For contents of previous issues see: www.ingentaconnect.com/content/berg/cftj/

Manuscripts

Articles for consideration should be submitted to *The Journal of Modern Craft*, c/o Glenn Adamson (email: g.adamson@vam.ac.uk). Manuscripts will be acknowledged by the editor and entered into the review process discussed below. Images should be submitted digitally, either on a disk mailed to: Glenn Adamson, Research Department, V&A, Cromwell Road, London SW7 2RL; or via a file transfer service such as yousendit.com or box.net, to Glenn Adamson (email: g.adamson@vam.ac.uk). Submission of a manuscript to the journal will be taken to imply that it is not being considered elsewhere for publication, and that if accepted for publication, it will not be published elsewhere, in the same form, in any language, without the consent of the editor and publisher. It is a condition of acceptance by the editor of a manuscript for publication that the publishers automatically acquire the copyright of the published article throughout the world. *The Journal of Modern Craft* does not pay authors for their manuscripts nor does it provide retyping, drawing, or mounting of illustrations.

Style

US spelling and mechanics are to be used. Authors are advised to consult *The Chicago Manual of Style* (16th Edition) as a guideline for style. *Webster's Dictionary* is our arbiter of spelling. We encourage the use of major subheadings and, where appropriate, second-level subheadings. Manuscripts submitted for consideration as an article must contain: a title page with the full title of the article, the author(s) name and address, a three-sentence biography for each author, and a 200-word abstract. Do not place the author's name on any other page of the manuscript.

Footnotes

Footnotes appear as "Notes" at the end of articles. Notes are to be numbered consecutively throughout the paper and are to be typed double-spaced at the end of the text. (Do not use any footnoting or end-noting programs that your software may offer as this text becomes irretrievably lost at the typesetting stage.)

References

Each article should include a list of references that is limited to, and inclusive of, all those publications actually cited in the text. References can be cited in one of two ways:

1) In the body of the text in parentheses with the author's last name, the year of original publication, and page number—e.g. (Rouch 1958: 45). In this case, titles and publication information appear as "References"

at the end of the article and should be listed alphabetically by author and chronologically for each author. Names of journals and publications should appear in full. Film and video information appears as "Filmography." References cited should be typed double-spaced on a separate page.

2) In end notes, following the formula:

Book: Author, *Title* (City: Publisher, Date), p. [page number(s)].
Article: Author, "Article Title," *Journal Title* [volume]/[number] (date), p. [page number(s)].

References not presented in one of the two above styles will be returned to the author for revision. Please refer to previous copies of the journal if in doubt about referencing style.

Figures

All illustrative material (drawings, maps, diagrams, and photographs) should be designated "Figures." They must be submitted in a form suitable for publication without redrawing. Drawings should be carefully done with black ink on either hard, white, smooth-surfaced board or good quality tracing paper. Computer-generated drawings of publishable quality can also be submitted, ideally saved as high-resolution TIF or maximum quality JPG files. High-resolution (600dpi or above) color or black and white photographs are encouraged by the publishers. Authors should indicate in the manuscript approximately where each figure should appear. All captions should be submitted as a separate Word document.

It is the author's sole responsibility to secure images, as well as worldwide rights to those images for both print and online publication. All reproduction costs charged by rights holders must be borne by the author. Any queries about the preparation or submission of artwork should be referred to the editors.

Criteria for Evaluation

The Journal of Modern Craft operates on a single peer-review basis, and sometimes seeks a second peer review in cases where multiple areas of expertise not covered by the editors are required. Articles will be reviewed anonymously by qualified academic reviewers, who are asked to consider originality of research; completeness of coverage; and assessed according to the following possible results:

1) Publish the article in its present form, with no substantial revisions
2) Publish with minor revisions. (In this case, final approval is subject only to review by the journal's editors.)
3) Publish with major revisions. (In this case, final approval is subject to a further round of external peer review, either by the same reviewer or a new reviewer.)
4) Decline to publish.

Detailed feedback will also be provided by the Journal's editors, both prior and subsequent to peer review, as appropriate. Keep in mind that this process may be time consuming, so it may some time for an article to be published following the initiation of the peer-review process. Ideally articles are published between 12 and 18 months following initial submission, but this is highly variable depending on the peer review process.

Editing

At the conclusion of the peer review process, the Journal's editors will work with you to complete a final draft of the text. This is submitted to Berg by the editors, approximately 6 months prior to the publication date of the article. Berg will then conduct a further copyedit of the text, layout the article pages, and send you a proof for final correction. This process is usually completed about 3 months prior to the publication date.

Offprints

On publication, authors will be sent a PDF of the final, published version of their article for personal use, and will be able to order a free copy of the issue in which their article appears. You will be contacted upon publication with details.